IEE Management of Technology
Series Editor: G. A. Montgomery

BUSINESS FOR ENGINEERS

Brian Twiss is an international authority on the management of technology. Experience in the management of advanced aerospace projects was followed by 20 years in academia, firstly at the Cranfield Institute of Technology and later at the University of Bradford Management Centre where he organised courses for senior technical managers. He has contributed to MBA programmes and run seminars in technological management in many universities and institutes throughout the world. He has undertaken consultancy assignments for major technological companies in the United Kingdom, Europe and elsewhere.

He is currently an independent consultant in technology management and forecasting, Reader in the Management of Technology at IMCB, and coordinator of the ESRC/SERC Research Initiative on the Successful Management of Technological Change.

Apart from his writing and teaching he has, in recent years, been increasingly involved in research including an MSC funded project on managing technical change. In 1985 he visited Japan to study management development for technological innovation.

The main emphasis of his work is the integration of technology and business from the MBA to top management levels, both for technical and non-technical managers.

He has written articles for many journals and is author or co-author of six books, one of which is a standard text in many universities and business schools throughout the world. He is a member of the editorial boards of the journals 'Product Innovation Management', 'R&D Management' and 'Research Policy'.

BUSINESS FOR ENGINEERS

Brian C Twiss

Peter Peregrinus Ltd on behalf of the Institution of Electrical Engineers

Other volumes in this series

Volume 1 Technologies and markets
          J. J. Verschuur
Volume 2 The business of electronic product development
          Fabian Monds
Volume 3 The marketing of technology
          C. G. Ryan
Volume 4 Marketing for engineers
          J. S. Bayliss
Volume 5 Microcomputers and marketing decisions
          L. A. Williams
Volume 6 Management for engineers
          D. L. Johnston
Volume 7 Perspectives on project management
          R. N. G. Burbridge (Editor)

Published by: Peter Peregrinus Ltd., London, United Kingdom

© **1988: Peter Peregrinus Ltd.**

All rights reserved. No part of this publication may be reproduced, stored in a retrieval system or transmitted in any form or by any means—electronic, mechanical, photocopying, recording or otherwise—without the prior written permission of the publisher.

While the author and the publishers believe that the information and guidance given in this work are correct, all parties must rely upon their own skill and judgment when making use of them. Neither the author nor the publishers assume any liability to anyone for any loss or damage caused by any error or omission in the work, whether such error or omission is the result of negligence or any other cause. Any and all such liability is disclaimed.

British Library Cataloguing in Publication Data

Twiss, Brian C. (Brian Charles), 1926-
  Business for engineers.
  1. Management—For engineering
  I. Title   II. Institution of Electrical
  Engineers   III. Series
  620'.0068

ISBN 0 86341 126 6

Printed in England by Short Run Press Ltd., Exeter

Contents

		Page
Foreword		ix
Introduction		xi

1 Engineering and corporate strategic management — 1
- 1.1 The changing environment of business — 3
- 1.2 Objectives, strategy and action — 6
  - A view of the future — 6
  - Strategic objectives — 8
  - Strategy formulation — 8
  - Decisions and actions — 11
- 1.3 Gap analysis — 12
- 1.4 Portfolio analysis — 15
- 1.5 Competitive strategy — 17
- 1.6 What business are we in? — 19
- 1.7 Response to outside pressures — 19
- 1.8 Is a strategic approach necessary? — 21
- 1.9 The role of the engineer — 23

2 Strategy for engineering and technology — 26
- 2.1 The need for a strategy — 26
- 2.2 Product or manufacturing investment? — 28
- 2.3 Horizontal and vertical integration — 31
- 2.4 The industry/technology life cycle — 31
- 2.5 Developing a technical strategy — 38
- 2.6 Status review — 39
- 2.7 Inputs to technical strategy formulation — 41
  - Corporate strategy — 41
  - Marketing and financial strategies — 42
  - Mandatory requirements — 43
  - Technology capture — 43
  - Technological opportunites and threats — 44
- 2.8 Competitive advantage — 47
- 2.9 Technical capabilities — 49
- 2.10 The risk profile — 50
- 2.11 Project proposals — 50
- 2.12 Techniques to aid strategy formulation — 51

	2.13	The form of the strategy	51
	2.14	The next stages	52

3 Technical progress and engineering decisions — 53
- 3.1 The need — 53
- 3.2 Historical background to technical progress — 55
- 3.3 The determinants of technical progress — 58
- 3.4 The elements of technology forecasting — 59
- 3.5 The inputs to forecasting — 62
- 3.6 The causal model — 65
- 3.7 Patterns of progress — 68
- 3.8 Technological substitution — 72
- 3.9 Attribute substitution — 74
- 3.10 Production substitution — 75
- 3.11 Forecasting techniques — 78

4 Engineering, marketing and product development — 79
- 4.1 The new product concept — 80
  - Creativity — 80
  - Technology capture — 82
  - Market knowledge — 83
  - Environmental forecasts — 84
  - The evolution of the product concept — 85
- 4.2 Product specification — 85
  - Product specification and the industry life cycle — 87
  - Attribute analysis and market segmentation — 87
- 4.3 Exploratory research and development — 90
- 4.4 Project evaluation — 92
- 4.5 Product design — 95
  - Design for manufacture — 96
  - Design for minimum cost — 97
  - Design for the operating system — 98
  - Design for market diversity — 98
- 4.6 Engineering and marketing integration — 100
- 4.7 Cost and pricing — 100
  - Price–volume relationships — 101
  - Unit cost and cumulative production (the experience curve) — 102
- 4.8 Product development — 104
- 4.9 Summary — 105

5 The engineer and production management — 106
- 5.1 The role of production — 107
- 5.2 One-off production — 108
  - Management of one-off production — 110
- 5.3 Repetitive manufacture — 113
  - Management of repetitive manufacture — 117
  - Management of process industry — 118
- 5.4 Types of production operation: summary — 119
- 5.5 Technological developments and production management — 121
  - The need for continuous change — 122
  - Flexibility — 123
  - Organisational integration — 123
  - The size and characteristics of the workforce — 126

		The changing role of the manufacturing manager	126
	5.6	Quality assurance	128
6		**Financial implications of engineering decisions**	**130**
	6.1	Capital budgeting	133
		Estimating accuracy	134
	6.2	Financial evaluation techniques	137
		Payback	137
		Benefit/cost	137
		Return on investment (ROI)	137
		Discounted cash flow (DCF)	138
		Risk and uncertainty	139
		Review	141
	6.3	Cash flow and project design	142
	6.4	Finance and engineering decisions	144
		Sunk costs	144
		Depreciation	146
		Fixed versus variable costs	147
		Capital intensity	148
		The experience curve	149
		Make or buy	150
	6.5	Value added	151
	6.6	Conclusions	152
7		**The management of technical change and technology transfer**	**154**
	7.1	The process of innovation	155
	7.2	Human barriers to change	156
		Lack of top management commitment	157
		The generation gap	157
		Premature critical evaluation	158
		Fear of change and risk aversion	158
		Self-imposed perceived constraints	160
	7.3	Evaluating a technical change programme	160
		What are the merits of the idea?	161
		How complex is it?	163
		How should it be implemented?	163
		What is at risk?	164
		How easily can it be understood?	165
		Summary	165
	7.4	Gaining support for technical change	166
	7.5	Organisational structure for technical change	168
		The hierarchical organisation	169
		New venture groups	170
		Matrix organisation	172
	7.6	Management for technology transfer	173
		The widening international technological base	173
		The escalating cost of R & D	173
		Rapid international communications	174
		The growth of international markets	175
		The growth of international production	176
		The growth of international R & D	176
	7.7	A corporate framework for technology transfer	177
	7.8	Conclusion	180

8		**The engineer: career progression and development needs**	**181**
	8.1	The changing nature of business	181
		The increasing corporate role of the engineer	182
		The global nature of the business environment	182
		The pace of change	183
		Closer integration of business functions	184
		The integration of technologies	186
	8.2	Engineering and business decisions: some differences	186
		Survival versus growth	187
		Quantitative versus qualitative information	187
		Importance versus priority	188
		Entrepreneurship versus management systems	189
		Individual, departmental and corporate values and objectives	190
		Risk versus security	192
		Literacy versus numeracy	193
		Cash flow versus profitability	194
		Summary	194
	8.3	The engineer's career development	196
		Phase 1: specialist	197
		Phase 2: junior engineering manager	197
		Phase 3: section leader	197
		Phase 4: project manager	199
		Phase 5: engineering director	199
	8.4	Career development and competences	201

Bibliography 203

Index 205

Foreword

Napoleon's taunting description of the British as 'a nation of shopkeepers' is no longer valid, if indeed it ever was; that is why this book by Brian Twiss is so important. Businesses in many countries are held back because some of those in charge know little about technology, and many of the engineers they employ know little about business strategy.

Most engineers have a sound knowledge of their own subject and of corresponding new and exciting possibilities. Enough of them have a grasp of technology in general and of the various technologies which impinge on their organisations. They still produce a great number of potentially important discoveries and inventions.

Why is it then that so many of us in the western countries leave it to others to apply technology in practice and to reap the financial and material benefits of what we have initiated? Is there a failure of communication between businessmen and engineers?

In small organisations, these two sorts of people may be members of the same tiny group, may even be a single person; and there is no problem. In larger outfits each profession must learn to listen to the others, and this book will help engineers at least to improve the communication link from their end – learning to think and talk a little more like shopkeepers! It may also give non-engineers some insight into what technology can do and how it can contribute to overall results.

Engineers know about technology; some are good inventors or designers, and the best of them know a great deal about managing technology and technologists. They may even understand some of the day-to-day principles of management in general. (Other books in this series, especially 'Management for Engineers' by D. L. Johnston, can help with this.)

What few engineers know enough about is business strategy, how to grasp the longer term threats and opportunities presented by technology and how to use such knowledge for the good of their organisations.

Few authors are more expert in this field than Brian Twiss. He has applied, in large and small organisations world-wide, the principles he presents here, and he has written and lectured extensively about them, as well as acting as a consultant to many internationally-known organisations.

Aegerter has drawn attention to the widespread though unreasonable expectation that technology can meet all human needs, not only material ones, at the press of a button. In the same way, those in control of enterprises sometimes believe that technology can be switched on instantaneously to benefit the operation (and incidentally switched off when cost savings are needed), without becoming an integral part of it. This situation must be corrected if we are to prosper; *that* is why this book is so important. I commend it to you – it may improve your career prospects (as a minimum just by making it easier for you to obtain a further relevant qualification), but better still it might really improve the future for us all.

G. A. Montgomerie 1988

Introduction

There are many reasons why people choose a career in engineering. But there is only one reason why engineers are paid to practise their profession. This is the belief that their efforts will lead to a better society by meeting the needs of either the individuals who buy the products they design and manufacture or the wider community through the purchases of central or local governments. No one is forced to buy an engineering product. They will only do so if it gives greater satisfaction than alternative uses for their money. Having made the decision to buy, they are still faced with the choice of the particular product. Thus the engineer and his actions can only be assessed in relation to money, markets and competition. Within the business his employer is not interested in the elegance of engineering solutions to problems; he is concerned with the profit contribution from satisfying a market. It is the consequences of engineering, not the nature of the engineering, which is his prime concern.

The engineer's professional training may have given him little preparation for his role in business. He will have spent a number of years acquiring an expertise in one or a few academic disciplines. His desire is to apply this knowledge in creating new products or solving technical problems. He aspires to technical excellence. He is not equipped to identify market opportunities or assess financial alternatives. Whilst recognising their importance he may regard them as the responsibility of others within the organisation. He may see himself either as the specialist who applies his skills to problems emanating elsewhere in the business or as the creator of opportunities which others should exploit. Thus we see two conflicting views of the role of the engineer in business. The engineer is motivated by the quality of his technical input, whereas the business is concerned with the relevance of his output.

It might be argued that this dichotomy is an over-simplification. However, there is a great deal of evidence to show that it does reflect the situation in many companies. It cannot be denied that a large number of industrial companies have gone out of business in the past 20 years or that many new products have failed in the market place. A major contribution to these failures is the way in which engineering skills have been employed within the business. There is a strategic dimension, where the company has failed to grasp the opportunities presented by new technologies or to recognise the threat they present to the existing business, for example the application of microelectronics in products previously based upon mechanical or electromechanical expertise. There are products for which the eventual market proves to be much smaller than expected or non-existent, such as the Concorde and the video telephone. And there are products which could have succeeded but for their costs, a consequence of design or manufacturing shortcomings, or poor quality. In very few of these cases does the fault lie with engineering competence; the failures are a consequence of the appropriateness and relevance of that competence to the needs of the business. For the most part engineering failures, such as the collapse of a bridge or the problems with the advanced passenger train, are relatively rare.

Examination of failures indicates some of the problems at their most dramatic. However, the symptoms can be detected in all companies since perfection is an ideal which few if any human activities achieve. Even in the most successful companies there is scope for improvement. It is the purpose of this book to explore ways in which the engineering contribution to business performance can be enhanced. Although directed to the engineer, a 'them' and 'us' attitude must be avoided; the aim must be to achieve an integrated approach where the engineer and the non-engineer make a combined effort to achieve a common objective. It is analogous with a marriage where each partner makes a distinct contribution to the creation, nourishment and well-being of the family. It will not succeed, however, unless there are shared values, mutual understanding and common objectives. There will also be a division of labour, and each partner will engage in activities which contribute to the performance of his or her particular role. Thus the engineer cannot ignore his need to develop his professional knowledge and competence; this is his unique contribution to the business which no one else can make. However, to ensure that this is harnessed to the needs of the business he must also play an active part in establishing corporate as well as technical objectives. Once these objectives are formulated they can only be achieved by an integration of effort across the company involving all functions. This requires an understanding of the operations of the company as a whole and the constraints within which other people have to work. Every engineering decision involves making a choice. The criteria by which these choices are made are not only technical but also involve finance, personnel, manufacturing and markets.

Although the close integration of engineering with the rest of the organisation has always been important, it has become even more so in recent years. There are two main reasons for this. Firstly, the business environment has become more turbulent, leading to greater uncertainty and an increase in the rate of change. This has brought with it a need for greater responsiveness and flexibility, which can only be satisfied where there are clear objectives, good communications and a unity of purpose. Secondly, the rapid advances in many technologies have increased their importance as a key resource in enabling a company to achieve a competitive advantage. Thus engineering or technology must assume a central role in the formulation of business strategies; this can only be accomplished with the active involvement of engineers who have an understanding of the capabilities and limitations of these advances together with an appreciation of the needs of the organisation. Thus there are two themes which run throughout all eight chapters of this book: the need to innovate, and the close integration of all business functions.

In order to fulfil its corporate role, engineering must discharge its responsibilities under three broad headings:

1 The effective application of technical knowledge in the design of products, processes and systems
2 The efficient management of the resources invested in the engineering departments to achieve the timely and cost effective achievement of technical objectives
3 The deployment of technology, in R & D, design and manufacture, to further the corporate objectives of the business.

These responsibilities can be termed technical, managerial and business, respectively. All are important. There will inevitably be some degree of overlap between them as will be seen later in this book, which primarily addresses the third.

This is not a book about business for an engineering readership. It is a book about engineering decisions and how they must reflect the needs of the business. It is not a book concerned with management techniques, although a number will be described. Systems can provide a framework but by themselves do not ensure that effective integration will ensue. This depends upon people throughout the business with different roles, knowledge, concerns and priorities. Thus the aim of this book is to give the reader an understanding of those with whom he must co-operate in order to further the interests of the business as a whole.

Chapter 1 describes some of the concerns of the top management of a business and discusses the main approaches to formulating a corporate strategy. The problems are seen to be complex, involving choices between conflicting claims for resources and their anticipated benefits in the short or long term. A coherent strategy is needed in order to avoid the threats and exploit the opportunities arising from the changes in the environment within which the business operates. Thus whilst the overall corporate objectives may remain constant for long periods, the means by which they are to be achieved – the strategy – must adapt to the business environment. The strategy, therefore is a mechanism for achieving timely and appropriate change. The chapter concludes with a discussion of the role of the engineer in the process of strategy formulation.

Chapter 2 shows how engineering activities can be co-ordinated with the business strategy through the development of a technical strategy. This also involves choices between the allocation of resources for long- or short-term developments of new products or processes, and horizontal or vertical integration. The industry life cycle concept is discussed and it is shown how the emphasis of technical investment needs to change to meet the needs of the business as it evolves. The impact of this on the status of engineering, the style of management and the managerial control systems is highlighted. It is concluded that there is no right managerial approach for all time but a choice between alternatives selected for their appropriateness to the stage of the life cycle. The process of technical strategy development is discussed together with a brief description of techniques.

Chapter 3 introduces the time dimension into technical decision-making. A brief historical review shows how different technologies have contributed to industrial growth over the past two centuries. It also indicates that we are currently experiencing a period where the application of the advanced technologies is leading to significant structural changes in industry. This underlines the increasing importance of technology as a factor in corporate decision-making. An examination of past trends indicates that they do not occur randomly; there are patterns of progress which can be identified. These patterns can be used to forecast the path of future advance. In most engineering assessments it is not difficult to identify what developments are likely to occur; it is much more difficult to form an opinion of when they will happen. The techniques of technology forecasting provide a valuable tool for introducing this time dimension into engineering decisions, many of which do not come to fruition for many years. The elements of technology forecasting and some of the most useful techniques are described briefly.

These first three chapters provide the framework within which the engineer must work. It gives a sense of direction but cannot ensure success, which depends upon the physical output from his efforts – new products or new processes. Their value to the business derives from the market place. Chapter 4 stresses the importance of a close integration between engineering and the market. Although the marketing department provides the formal interface with the market it is stressed that the engineer needs to develop a sensitivity to customer needs in addition to maintaining a close relationship with his marketing colleagues. The product itself is only the mechanism whereby the potential of technology is used to satisfy customer needs. This chapter discusses how this can be achieved through the successive stages of new product concept derivation, product specification, project evaluation and product design.

Chapter 5 is concerned with production. The engineer is involved mainly at two levels. If he is working in R & D or design his new products must lend themselves to cost-effective manufacture or his new processes to improved manufacture. If he is employed in production he must ensure that his processes and systems exploit the potential of technology to the full. In either case a close integration is essential. A sequential process starting with design and passing to manufacture and then to marketing is seen to lead to delays and additional costs. All three activities must be closely associated at all stages.

It is noted at the beginning of the chapter that manufacturing is undergoing a transformation. Whilst many engineering companies still rely heavily on traditional practices an increasing number are adopting the new microelectronics-based techniques such as CAD/CAM, CIM and AMT. Both types of system are described and contrasted. It is seen that many of the former concerns of manufacturing managers with inventory control and production scheduling and control are becoming of decreasing importance. In their place there is need for a new approach to manufacturing management to reflect the demands of continuous change, flexibility, organisational integration, the changing size and characteristics of the workforce and the role of the production manager.

Chapter 6 considers some of the financial implications of engineering decisions. The resources devoted to engineering represent an investment from which the business aims to obtain a financial return. It is beyond the scope of this book to enter into a detailed account of accounting and financial practices. It concentrates upon two main considerations that the engineer must bear in mind. Firstly, he must draw a clear distinction between the needs of the accounting conventions and his use of financial data in making his decisions. Every decision he takes has a financial implication. However, if he chooses inappropriate accounting data for making his decisions these may not be the best from the point of view of company profitability. A number of these considerations are discussed. Secondly, he must not forget that the ultimate aim is profitability not cost minimisation. This raises a number of problems for the engineer. Cost is tangible whereas the size of the resultant profit is uncertain. Corporate cash flow constraints ensure that all expenditure is examined critically. Strict cost control is essential in order to eliminate waste. Nevertheless, the engineer should not be so obsessed by the needs of cost reduction that he takes decisions that are not in the interests of the long-term objective of the company, namely profit.

Chapter 7 deals with two topics – the management of technical change and technology transfer. The engineer is a major change agent in the business, for he alone can assess the potential business value of new technology. It is seen that many of the most successful innovations have resulted from the entrepreneurial activities of individual engineers in the face of opposition elsewhere in the company. The process whereby change is effected is discussed. The most common human barriers to change are then examined. It is seen that many of the objections raised to any proposal to innovate, whilst they may seem undesirable, reflect real threats to others in the business. A system for maximising the chances of success is then described. It has to be recognised that it may be necessary to adapt a proposal in order to make it more acceptable. The importance of the engineer's entrepreneurial role is becoming more widely recognised and is reflected in organisational structures to facilitate it; these are discussed.

No business can be self-sufficient in technology. This is of increased relevance as the environment for technology is changing. A number of current trends are discussed. These include the widening international technological base, the escalating cost of R & D, rapid international communications, and the growth of international markets, production and R & D. A corporate framework for assessing these trends and examining business alternatives to transferring technology is described.

Chapter 8 draws together a number of threads running through the previous discussions. It commences by reviewing the changing nature of business, stressing the increasing role of the engineer, the global nature of the business environment, the rapid pace of change, the need for closer integration and flexibility within the business, and the integration and convergence of technologies. It then examines some of the differences between engineering and business decisions and, in particular, the conflicting criteria which the business decision-maker has to resolve. The complexity and qualitative elements of many business decisions can lead to poor communication with the engineer whose training emphasises the importance of numeracy and quantitative criteria. Finally the changing requirements as the engineer's career progresses from specialist to engineering manager, section leader, project manager and engineering director are discussed in relation to technical and non-technical competences. Within the technical field he needs to develop skills in additional disciplines and a knowledge of advances in the technical environment. The non-technical competences cover man management, knowledge of other functions and the business environment. These competences have to be gained continuously throughout his career.

The purpose of this book is to set these requirements in the total business context. It is maintained that this is a process that the engineer must address from an early stage in his career. The competences he needs must be developed by efforts to prepare himself for the challenges of the future. Their acquisition cannot be left until they are required; by then it will be too late to develop the depth of knowledge and understanding that will be needed.

Because of the wide scope of the coverage of this book detailed references have been omitted. The bibliography has been carefully selected to provide the basis for further study.

In conclusion it must be stressed again that the business of engineers is business. This responsibility cannot be discharged by allowing the management of technology to go by default. Once the engineer has familarised himself with the complexities of the wider business world he will find in them an intellectual challenge, different but equally satisfying to that which he gains from employing his technical knowledge.

Chapter 1
Engineering and corporate strategic management

All engineers are employed to further the interests of the organisations within which they work. This applies equally whether they are engaged in designing or manufacturing commercial products in private industry or in non-profit organisations such as government agencies or public utilities. This book is written primarily for those employed in the private sector where profitability provides a simple measure of the organisation's success. The concepts described in this book are, in the main, equally appropriate to those in non-profit organisations, although their application may often present problems where the objectives are likely to be less clearly articulated. In either case the role of the engineer is to add value.

The overall objectives for business management, as described in the textbooks, are to ensure that the organisation survives in the short term and grows in profitability in the longer term. Although this is a truism which cannot be denied, a brief examination of these objectives reveals many of the problems besetting managers at all levels. Only rarely will these two objectives – survival and growth – be compatible. Survival will frequently demand the diversion of limited resources from the investment required to ensure long-term growth. Thus a conflict arises between the two objectives which management must resolve.

How can management resolve this conflict between these rival claims for resources? There are no simple answers, since any diversion of investment funds for the future, whilst strengthening the company's ability to overcome short-term problems, inevitably involves some sacrifice of growth aspirations. Thus a balance has to be struck. In coming to its decision management will be accepting a degree of risk in the present for benefits anticipated for the future. Some element of uncertainty will always remain. Furthermore, no sophisticated calculations can tell us what is the appropriate level since the risk accepted is a reflection of the attitudes of the management involved, whether they are risk averse or risk takers.

We can already begin to comprehend the complexity of the decision-making environment of business managers. Four elements are worthy of comment:

There are no right answers Later we shall examine a wide range of conflicting considerations which need to be balanced, of which survival versus growth is but one. Although there is a wide and growing number of management techniques which help us to intellectualise the problems, they do not provide the answers; they only assist. In the final analysis we have to rely upon managerial judgement, albeit operating within bounds set by rational analytical processes.

Risk and uncertainty cannot be avoided Risk is a consequence of uncertainty, which arises from two main causes. Firstly, the complexity of the factors involved in the majority of managerial decisions is such that all the relationships are imperfectly understood; it is also unlikely that all the information which might be desired for decision-making is either available or accurate. Secondly, business is operating in an external environment which will always exhibit behaviour which cannot be fully anticipated, for example violent fluctuations in oil prices or exchange rates.

Human values are reflected in the internal culture of the business In recent years the importance of corporate culture has been receiving increased attention in the literature of business. An examination of any two companies competing in the same market (e.g. Unilever and Procter and Gamble; Sainsbury and Tesco) indicates that they have individual characters which reflect different corporate value systems. Thus in a given situation they will respond differently. This will manifest itself in many ways, such as in their attitudes to risk, quality and service. Although some sets of values might be generally regarded as 'bad' there are many 'good' sets which are equally valid although different. For example, one company might contemplate any decision provided that it is within the letter of the law, whereas another will pay much greater heed to the spirit of the law. Few companies attempt to define their corporate culture explicitly, yet all decisions including those in engineering must be within the bounds set by this culture.

Profitability is the measure of business success It is only in financial terms that all the resources used in a business – manpower, materials etc. – can be measured by a common unit. This means that all managerial decisions, including those in engineering, must be justifiable and ultimately expressed in financial terms. Only thus can they be aggregated for general management purposes. It is often stated that the purpose of business is to make profit. This may be the general view of the financial community, although most engineers would widen this statement by saying that the aim is to make profit through the sale of products incorporating good design features and efficient manufacturing processes. Some would also argue that many businesses have social as well as purely profit motives. Nevertheless, profit must remain the prime consideration. Two observations in respect of engineering decisions should be made here. Firstly, engineers must at all times consider the financial implications of their decisions and have, at least, some rudimentary financial and accounting understanding. Secondly, cost reduction, although of great importance, is not the prime consideration; there are occasions when cost minimisation is not synonymous with profit maximisation.

Good engineering and good management have a great deal in common. They both combine a scientific analytical approach with pragmatism. They both require creativity, good judgement, initiative and enthusiasm. In spite of this their emphasis is different, for the engineer educated in the scientific tradition is seeking for the optimum if not the ideal solution to his problems. He will regard risk as something to be eliminated rather than something to be lived with. The aim is to establish the facts; judgement is regarded as an undesirable although necessary evil. Ambiguity is to be avoided. The manager, however, is operating in a less certain world where analytical techniques are helpful but secondary to good judgement. Some degree of risk is inseparable from all his actions taken in an uncertain world. Although ambiguity is undesirable, it is often unavoidable.

Generalisations such as those stated above can never tell the whole story. Nevertheless it is important for the engineer to recognise that there is a difference in emphasis necessitated by the different and wider environment in which the manager operates. This may help to explain some managerial decisions which might otherwise appear inexplicable. The reader is asked to bear these differences of emphasis in mind when reading what follows.

1.1 The changing environment of business

The successful business adapts its strategies and policies to the environment in which it has to conduct its affairs. This environment consists of four elements which are continually changing – the economic, the social, the political and the technological. If the environment changes, so must the organisation. But change takes time to work itself through a company for a number of reasons, however dynamic the management may be. There is a great deal of inertia in any organisation, much of it unavoidable. For example, the physical facilities and their equipment represent a significant investment made in the past which cannot be easily or economically replaced. More importantly, human beings are resistant to change, particularly when it is thought to affect their own personal position or make their skills irrelevant in the future. Even where there is no overt opposition it still takes a considerable time to change attitudes, organisational structures and work methods. New technologies will often require the acquisition of new skills either through retraining or all too frequently through recruitment of those who possess them in place of older employees. For these reasons an organisation needs an early warning system so that it can prepare in advance for the changes which will ultimately be forced upon it if it is to survive. Thus the style of management should be *proactive*, that is anticipating and planning for change, rather than *reactive*, where change has to be introduced to meet competitors' threats which have already begun to erode the company's position. Increasingly the management of change is being recognised as an essential feature of the successful business.

4 Engineering and corporate strategic management

It has already been noted that an organisation has to adapt itself to the general environment within which it operates. The influence of changes in the business environment can be seen to have a profound impact on management style and those aspects which are given greatest managerial attention at any time. It is, therefore, instructive to examine briefly some of the major characteristics of the business environment over the past 25 years and note the effects they have had upon the major concerns of management. This examination is considered in decade intervals as a simplification; in reality most changes evolve over time and it is not always easy for management to recognise what is happening whilst it is occurring. Nevertheless it is sensitivity to the business environment in the present and to what is likely to occur in the future which distinguishes the successful companies and enables them to adopt a proactive stance.

The post-war period to the 1960s was marked by steady expansion in the developed economies. Thus the external environment for most companies exhibited relative stability with the promise of continued growth. This assumption was rarely questioned. It posed few threats and presented relatively few uncertainties. Since this state of affairs was taken for granted business success was largely equated with how well a company managed its own internal affairs. It was during this period that formal corporate planning was introduced into many large companies on the assumption that precise future targets could be established and would be achieved with a high degree of confidence, provided of course that the right actions were taken within the plan. This approach was highly deterministic and was based upon the assumption that the future would continue to exhibit the same characteristics of steady growth. Forecasting, for example, was based upon extrapolation of past trends and there was an implied belief that these forecasts were soundly based predictions.

There was also a great belief in the power of quantification particularly when based upon the techniques of operational research (OR). One must accept that the majority of firms had made little progress in the application of these planning and OR techniques; nevertheless it was accepted that this was the direction in which advanced management was evolving. In some ways the approach was mechanistic even when applied to human management in, for example, the rapid growth of techniques such as management by objectives.

The relatively threat-free environment was also reflected in financial policies. Where future uncertainties were regarded as slight it was possible to take decisions in order to sustain growth which at another time might be regarded as risky. As a consequence the enterprising firm was considered to be one that used its debt capacity to the full to gear up its balance sheet in order to invest for future growth. Survival was not seen as a problem and the emphasis was placed upon exploitation of opportunities for growth.

Although there were signs in the early 1970s that the growth rate could not be sustained indefinitely, this had little impact on business thinking until the shock of the first oil crisis in 1973. The years following the oil crisis marked the beginning of an age of discontinuity. The external environment became all-important and changes in it were promoted to a central concern in top management thinking. One began to hear much more of corporate strategy as a concept where environmental analysis played an important part. Survival and conservative financial policies became dominant. Many of the companies which had incurred a heavy burden of debt finance faced economic difficulties that they were unable to survive. As a consequence this period was characterised by an emphasis on cash management and cost reduction. There was rationalisation of product lines and business units. Portfolio analysis (e.g. Boston Consulting Group Matrix) was popular, with a focus on identifying activities which should be discarded. Because of the uncertainties attention was focused on short-payback activities and a reluctance to support innovations, particularly if their benefits were of a long-term nature.

During the 1980s there has been another shift in emphasis. There is a growing realisation that, although there can be no return to the stability of the 1960s, management must think of the long term in spite of the many uncertainties which cannot be eliminated. The concern is with strategic management and the examination of alternative futures within an analytical framework. Thus long-term forecasting makes use of scenarios in which extrapolated trends, whilst remaining an important input to the activity, form only one of the approaches used.

The emergence of new technologies, particularly microelectronics, is having an increasing impact on both products and manufacturing process. This period could be characterised as one of technology-driven change. Associated with this is a growing commitment to innovation, entrepreneurship and new business ventures. Whilst organisations are still planning for the longer term, there is an acceptance that they must be more responsive to unforeseen changes in the business environment. This can only be achieved by flexible, adaptable smaller organisations. Wherever possible the large monolithic companies based upon the assumed benefits of economies of scale are being broken down into units that are smaller, more responsive and more easily managed. This has important implications for engineers. Technology is being promoted to the centre of the corporate stage. One might question whether this can be accomplished without the active involvement of engineers or technologists at the highest levels of business management. But in order to fulfil this vital role the individual engineer must take a wider interest in the business as a whole than has been usual in the past. His horizons must be enlarged and he must prepare himself for the wider role of general management.

It should be noted that this brief historical sketch of the past 40 years shows both an evolution in management thinking and a change of emphasis to meet the circumstances of the time. We can be confident that in the decades ahead the thinking will continue to evolve and there will be further changes in the external environment. The successful company will be that which keeps abreast of and applies the most advanced managerial approaches and, at the same time, maintains harmony with its environment.

6   Engineering and corporate strategic management

In the following sections of this chapter the most important concepts of corporate strategic management will be described. They should be regarded not as alternatives but as complements. Although they may be added to in the future, the concepts themselves will remain valid. What will change, however, is the relative importance of the factors to be considered in the analyses.

1.2 Objectives, strategy and action

In order to pursue a purposeful path into the future, three key elements are required:

Objectives   These should be clearly stated, explicit and (wherever possible) quantified statements of where the organisation aims to be at some time in the future.

Strategy   This is the path which the organisation intends to follow in order to reach its objectives. It will be selected from a number of alternatives to give the best balance between the company's abilities and its environment.

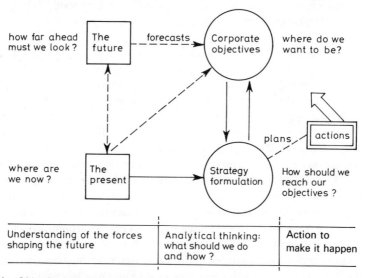

Fig. 1.1   *Objectives, strategy and action.*

Decisions and action   These are what the managers actually do in order to ensure that progress is made towards the objectives.

These three key elements are shown diagrammatically in Fig. 1.1.

A view of the future
In Fig. 1.1 'the future' appears as some given date, whereas in reality it extends indefinitely forward from tomorrow. However, it is essential to set one's sights

on some long-term target if the company is to adapt itself in advance to future needs. In the absence of a long-term view, decisions will become incremental and vulnerable to change. This raises the question of how far ahead it is necessary to look. There is no clear answer to this since it will depend upon the characteristics of the company and its industry. In general, however, three factors will need to be considered:

Decision lifetime The critical factor here is usually the nature and life of capital equipment. A decision to construct a chemical plant or a power station implies that there will be an economic justification for it throughout its working life, which could extend to 20–30 years ahead. Thus such decisions require a long-term forecast associated with a clear long-term objective for the organisation. For it implies that the company building the chemical plant takes the view not only that the plant will be employed profitably throughout the period but also that there is a high probability that the company will wish to remain engaged in the manufacture of chemicals.

Flexibility In the examples given above there was little flexibility. If you build an ethylene plant it can only produce ethylene. Should a replacement for ethylene emerge in its major uses the plant will be of little value; indeed it may be of negative value if it is expensive to remove. Only the technologist is capable of assessing the potential threats from new technology. On the other hand, many organisations are much more flexible and able to respond more quickly to outside changes as they occur. If a company is labour intensive, workers can be shed (albeit at a cost) in a relatively short time. If its equipment has a resale value or can be used for a wide product range, as is the case of general purpose machine tools, it can also adapt itself quickly. In general, although not always, smaller companies are likely to be more flexible than large.

Development time This can cover both technological and new business development. Technological development time falls into two main categories. In some industries product development is a lengthy process. The development of a new aircraft, engine, nuclear power plant or drug can cover upwards of ten years. Thus it is necessary to establish long-term technical specifications to ensure that it is competitive at the time of launch and for a considerable time afterwards. In other industries it is necessary to engage in long-term research where the existing technology is likely to be exposed to competition or indeed substitution by a new technology. A good example of this is provided by the Japanese electronic companies, which devote considerable resources to taking a 25–30 year view of technology; they engage in basic research in, for example, biotechnology in the belief that biological computers will eventually replace electronic. In a few large industries it may be necessary to contemplate long-term business diversification. This is a problem facing the major oil companies, which contemplate the reduction of oil supplies in the next century. A major strategic decision for the future has to be made well in advance to identify and invest in other activities, which must have a significant growth potential if they are to reach a substantial size by the time they are required.

Such considerations should lead many companies to consider more deeply the implications for their businesses of trends over a longer term than the usual 5 to 10 years. This indicates the need for forecasting, particularly for technology, which will be discussed in a later chapter. It must be recognised, however, that such forecasts cannot be expected to be accurate. Nevertheless, there is no way of avoiding the problem that a view of the future has to be taken if the characteristics of the industry or the company make it unavoidable.

Strategic objectives

Once a view of the future has been taken it is necessary to set realistic strategic objectives for the organisation. Furthermore these objectives must be established with sufficient clarity to enable decisions to be taken. This raises some difficult issues; as time progresses so may new information cause the view of the future to be changed, bringing about the need to modify the objectives. This is unavoidable and few if any companies will ever achieve the objectives set many years previously. Nevertheless it does provide a framework for decision-making. The penalty for inappropriate objectives will inevitably involve considerable financial loss and sometimes the failure of the company. But where the nature of the business is inflexible, this is a risk which has to be taken; no managerial approach can guarantee survival. As has been stated before, risk is inseparable from business; all that can be hoped for is that the risk will be reduced by the establishment of objectives set following a rigorous analysis of the best information available regarding the future. Where the uncertainties are deemed to be high a company might consider strategies to increase its flexibility. This may take various forms such as joint development with another manufacturer, or the increased use of subcontracting.

Strategy formulation

There will always be a variety of strategies by which a set of objectives may be achieved. In formulating the strategy which the business will adopt a balance has to be struck between the threats and opportunities revealed by the environmental analysis, and the organisation's ability to meet the threat or exploit the opportunities. This analysis is frequently referred to as SWOT (strength and weaknesses; opportunities and threats). However, great care needs to be exercised in assessing the strengths and weaknesses since it is a common failing to consider a particular factor as a strength or a weakness without sufficient thought. For example, is it a strength or a weakness that a high proportion of output is bought by one customer? Most students will state that this is a weakness because of the vulnerability of the business to the loss of that single customer. However, analysis of successful companies shows that many of them have been in that position for many years, as evidenced by the growth of some major suppliers to Marks and Spencer. The problem arises from the analyst's assumption of what is good or bad; these assumptions must always be questioned since they are generalisations which may not be true of the particular situations being examined.

A further complication arises from the fact that a feature of the company's current strength may be a weakness in relation to the needs of the future. In recent years many companies with an expertise in precision mechanical engineering (e.g. the Swiss watch industry) have found that this past strength is irrelevant to a future based upon microelectronics, a technology in which they may have no capability. Thus the simple analysis may indicate that it is necessary to build up an electronic expertise, but this is more easily said than accomplished. The strategic analysis can be seen to be merely the starting point for a major managerial reorientation to bring about the necessary changes. Furthermore, these changes may affect the whole way the organisation conducts its business. Returning to the watch example, a company may have strengths in its retailing and repair activities which are of little value when the technology of the product changes; today electronic watches are sold in supermarkets rather than specialist shops, and the product is rarely repaired.

The points considered above can be summarised in a number of questions:

1 What are the current strengths and weaknesses of the organisation?
2 Is the assessment of a factor as a strength or weakness based upon assumptions which may not be valid in a particular situation?
3 What are the capabilities required for the future?
4 Are current strengths relevant to the future?
5 What is required to overcome the weaknesses in relation to future needs?
6 How can these weaknesses be overcome within the company and change introduced?
7 What will be the wider impact of these changes throughout the operations of the business?

Although the SWOT analysis covers the whole of a company's operations it is evident that the technological and engineering capabilities are a central consideration, and they will be considered in greater depth in Chapter 2. It must be stressed, however, that the impact of changing technology will in many cases have a profound affect in other areas of the business in which the engineer has not previously had a major involvement. This is particularly true of the impact of the many applications of microelectronics in manufacturing, which often require a new approach to the total organisation. As a consequence it is becoming increasingly necessary to regard a company as an *integrated whole* in which the individual engineer cannot play his full part unless he is educated and motivated to take a wider interest than has been traditional in the past.

The assessment of threats and opportunities must relate to both the present and the future, although the latter is the prime consideration in strategy formulation. The main components of the business environment are normally analysed under the following headings: technological, economic, social and political. Stated in this broad way, a full analysis should consider the totality of all trends throughout the world. This, however, is impossible to achieve in practice, so judgement has to be used to decide what factors need to be analysed in some detail. In exercising this judgement, however, it must be recognised that the scope of the business environment has been growing throughout the past two centuries; there are now few companies which are not influenced by international competition, even when they are not involved in international operations themselves. Thus the initial scanning of the environment must be wide ranging even if it is superficial. In retrospect it can be seen that few of the threats which have brought about the downfall of major companies could not have been

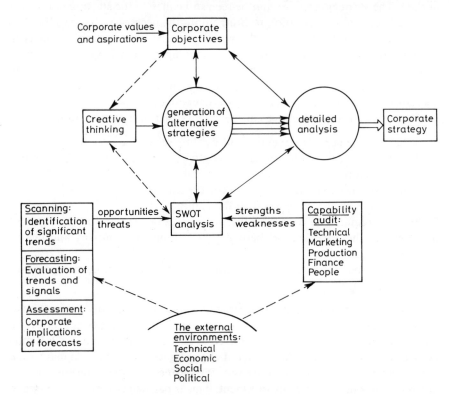

Fig. 1.2 *Elements of strategy formulation: an iterative process.*

foreseen. The difficulty lies in spotting these threats in advance and assessing their importance. The problems in doing this must be recognised, but the threats will never be spotted if management's field of view is too restricted.

Having established those significant factors which could have a bearing on the future direction the organisation should take, it is then necessary to carry out detailed forecasts. These forecasts will aim to give a quantitative assessment of the value of the phenomena studied through time. The most important techniques for forecasting, particularly in relation to technological change, will be explained at greater length in Chapter 3. At this stage in our discussion it is sufficient to recognise that the assessment of the future should be carried out in three broad stages:

Environmental scanning To identify those economic, technological, social and political factors, in a wide geographical context, which might be significant in the future.
Detailed forecasting To quantify these significant factors through time.
Strategic implications To assess whether these factors represent a threat or an opportunity, quantified so far as is possible.

The elements we have discussed for strategy formulation are shown diagramatically in Fig. 1.2.

Decisions and actions
The process of setting objectives and formulating a strategy is only a means to an end. It is analogous to planning a car journey where the destination is known and the route is selected with the aid of an imperfect map. But in order to reach the destination and at the right time it is necessary to have the appropriate vehicle in good running order; there must be fuel available; and the driver must be competent, able to read the map and capable of coping with unforeseen problems. So it is with running a business. The strategy must be translated into operating policies within which appropriate decisions are made and action is taken to ensure that the decisions are put into effect. Furthermore, management must be capable of taking action in the light of unforeseen difficulties and where necessary modifying the objectives and strategy in the light of new information. Even this is not sufficient, since managers are reliant upon others to implement their decisions. Thus the successful organisation must achieve a synthesis of *thought, action* and *motivating others*.

Many large companies with sophisticated strategic planning operations have been criticised for their lack of achievement. This argument was given added force by the publication of the book *In Search of Excellence: Lessons from America's Best-Run Companies* in 1982. The authors, Peters and Waterman, examined 62 successful US companies from which they noted eight attributes for excellence, namely:

1 A bias for action
2 Close to the customer
3 Autonomy and entrepreneurship
4 Productivity through people
5 Hands-on, value driven
6 Stick to the knitting
7 Simple form, lean staff

8 Simultaneous loose/tight properties, i.e. 'fanatic centralists group around the few core values they hold dear' but 'have pushed autonomy down to shop floor or product development team'.

It should be noted that this sample included 15 high-technology firms, amongst which were Hewlett Packard, IBM, Texas Instruments and Xerox.

Since the publication of *In Search of Excellence* there has been a major debate regarding the relative virtues of the 'old values' and the concepts of strategic management. In sharp contrast is the approach of major Japanese companies such as Hitachi, Matsushita and Mitsubishi which devote a considerable effort to establishing their long-term objectives. Hitachi, for example, has a think tank of 80 people, half technologists and half economists, concerned with a 20–30 year time frame. Once the long-term objectives are established, financial resources are made available with a minimum of financial evaluation of individual projects. This investment is maintained in the interests of long-term growth irrespective of the short-term economic climate.

Thus we see what appear to be three different philosophies for running a business. In the first, the emphasis is placed on setting long-term objectives and achieving them almost regardless of cost. In the second there is a concentration on strategic analysis. And in the third the focus is upon managerial action. This is undoubtedly an over-simplified picture since all three approaches are present to some extent in all successful companies. One might conclude that these should be regarded not as alternatives but as three essential elements in the effective management of a business. The appropriate mix of emphasis will depend upon the characteristics of the individual company and the industry. To a large extent this is dictated by the nature of the technology, engineering development lead times and the type of manufacturing processes.

1.3 Gap analysis

Another approach frequently used in strategic planning is that of gap analysis. A major element of any planning technique must relate to the financial performance of the company expressed in terms of either turnover or profit. Since profit is the ultimate prime objective for the organisation an analysis based on profit rather than turnover is likely to be the most useful in determining corporate policies. The gap is expressed as the difference between the corporate aspirations for profit and the expected achievement in the absence of new activities (Fig. 1.3).

At any time the company will have a portfolio of activities. Some of these are likely to be in decline, some will be in the early part of their growth and others are likely to continue at their present rate for the foreseeable future. In an organisation which has made inadequate plans for the future there will be a gap which needs to be filled by new activities if the financial objective is to be achieved. These activities fall into one of three categories – new products, new markets for existing products, or new businesses. Thus the gap gives an indication of the degree of innovation required.

Although the gap analysis appears to give a simple representation of the needs, problems can occur in the practical application of the technique. In order to understand these problems, one must consider how the gap has been derived and used. There are four stages, as follows.

The profile for the future desired profit: This represents the aspirations of top management which, if they are to provide a useful planning basis, must be realistic. Rarely does one see a curve sloping downwards, although one knows that there are times in a company's history where a temporary reverse in profit growth is unavoidable. It might also be asked whether a level which reveals no gap does not reflect a lack of innovative ambition. Thus there is no 'right' value for this curve since it must be a true reflection of top management's aspirations. However, if it is to be a useful guide the level set must be sufficiently ambitious to act as a spur for innovation, but not to such an extent that it loses credibility by becoming unrealistically optimistic.

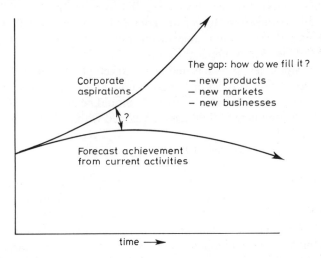

Fig. 1.3 *Gap analysis.*

Forecasts for activity performance: The analysis can be carried out using several bases for the contribution to the projected performance. Fig. 1.3 has been drawn using the contribution from the existing product portfolio. It could equally be drawn for market sectors, geographical areas or operating divisions. In theory, the aggregate should be the same whatever the basis used, although in practice there is likely to be some variation. It must be recognised, however, that these projections are forecasts and as such are prone to the uncertainties inherent in all forecasting.

Agreement on the size of the gap: The size of the gap would appear to follow automatically from the drawing of the two curves. However, in the author's experience there is not likely to be a ready acceptance of this when a substantial gap is revealed. This is a consequence of the indication that the corporate financial objectives are unattainable without significant new activities. The response of the managers involved will be a reflection of their own attitudes towards innovation. Thus in an organisation which is reluctant to embrace major change there will be a tendency to argue for a reduction in the size of the gap by reviewing the forecasts or reducing the financial objectives.

It can be seen that for the engineer the project portfolio gap analysis is an indication of the magnitude of the task he faces for the future. It establishes the technical objective in financial terms and can be used to arrive at the budgets for research and development and investment in new manufacturing processes; this aspect will be explained further in Chapter 2.

The filling of the gap: At the corporate level the debate will centre upon how the gap should be filled in the aggregate rather than in detail. Thus there will be different views about the respective contributions that can be made by new products, new markets or new businesses. To some extent this will reflect the culture of the organisation. In general companies can be broadly categorised into two classes; those that aim to grow through organic growth of the existing business, and those that prefer acquisition. In the former category the emphasis will be placed upon new products and new markets. In the latter it may favour acquisitions to achieve vertical or horizontal integration or alternatively diversification into entirely new business areas. It should be noted that new businesses can be based upon new technologies developed within the existing corporate structure; this is a growing area for corporate diversification and is another reflection of the increased role of technology in business.

Whatever the corporate preference some contribution must be made by new products and new markets, with a high probability that in most organisations some consideration must be given to new businesses. Within the board of directors there will be differences of opinion about where the emphasis should be placed. For example, the engineering director is likely to favour new investment in technology and manufacturing processes, whereas the marketing director is likely to favour an increase in the marketing budget. Where the balance lies must be a matter of judgement following detailed discussion with a degree of compromise. This raises an important point which will be discussed later, namely the ability of the engineer to persuade his colleagues. Since the agreed level of support for engineering investment is a matter of judgement and negotiation in the absence of any absolute values, it will inevitably depend to a large extent upon the communication skills and persuasiveness of those involved. Although this may be regarded as undesirable it is a feature of the real world in which many engineers exhibit serious weaknesses.

In spite of the problems associated with its application, gap analysis is a valuable tool in assisting corporate decision-makers. Like so many management techniques it helps to make the considerations explicit and provides a framework for informed discussion leading to decisions.

1.4 Portfolio analysis

Every company is an amalgam of a variety of activities, be they products, markets or business areas. They will vary in their current performance and in their potential to earn future profits. A number of visual representations in matrix form have been developed in order to assist corporate management in assessing which areas should be supported and which should be de-emphasised or abandoned.

The original matrix presentation was developed by the Boston Consulting Group (BCG) and analyses a company's portfolio of activities in terms of their

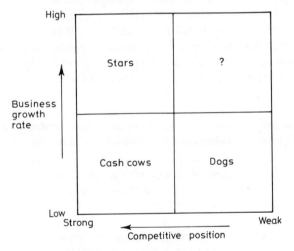

Fig. 1.4 *Portfolio analysis: Boston Consulting Group matrix.*

growth rates and the firm's competitive position (Fig. 1.4). In this 2 × 2 matrix the activities are categorised under four headings as follows.

Dogs: These are products or businesses in which the company has a relatively low competitive position (or market share) and where the overall growth rate is poor. Usually it is extremely difficult to improve the position of dogs without major injections of finance which, in view of the poor potential growth rate, would be hard to justify. Normally it is considered desirable to divest such activities and invest the cash generated in areas of the business with better prospects.

Question marks: This category of activity is one where the company has a weak position in a business with a better than average growth prospect. This presents a difficult problem, for the prospects for the overall market are good although the company has a relatively poor competitive position. There is an axiom that one cannot stand still and that the choice is always between growth or decline. Growth requires a considerable injection of cash to convert the question mark into a star. Whether this can be accomplished will depend upon the characteristics of the industry and the ability of the company. For example, in the computer industry many companies have tried to challenge the position of IBM, sometimes with products considered technically superior, but with notable lack of success. If the finance is not available, or the prospects of success are insufficiently good, it is usually preferable to abandon the question marks.

Cash cows: These are activities where the company has a large market share, albeit in a mature market with low growth potential. The name 'cash cow' derives from the fact that little new investment is needed to maintain the position of high market share or even dominance. In some cases they will be yesterday's products approaching eventual decline; in other cases they may face few immediate market threats but profit margins will be low. (The distinction between cash flow and profit will be examined in greater detail in Chapter 6.) Usually it will be found that it is the cash cows which make the major contribution to turnover, employ most of the workforce and utilise most of the physical facilities.

Stars: These are tomorrow's products or businesses where the company has a strong position in a growing market, although at the time this market may yet be small. These are the activities which normally require large cash injections in order to maintain their position, although they may be highly profitable. The natural progression will be from star to cash cow and perhaps from there to dog. Considerations of cash flow usually dictate a corporate balance between stars and cash cows.

There have been a number of developments of the simple BCG matrix. Descriptions of these will be found in the literature of strategic management. One widely used matrix is a 3 × 3 grid combining competitive position and industry attractiveness (Fig. 1.5). As with all techniques these matrices must be used with care, but they have the great advantage of presenting the total portfolio of activities in an easily understood visual display. It must be recognised that each of the axes of the matrix combines a number of factors. Competitive position, for example, represents more than market share, which can be easily quantified. Furthermore, the matrix will usually be drawn up in the light of the current position and represents a snapshot rather than a moving picture. In particular these matrices are useful in highlighting the dogs, and can play an important part in reaching a decision to divest. They also indicate the areas for future growth, although the greater uncertainties surrounding new business areas present greater problems for the decision-maker. Their value is indicated by their use in such major companies as Shell, Proctor and Gamble, AKZO, Ciba-Geigy and Kodak.

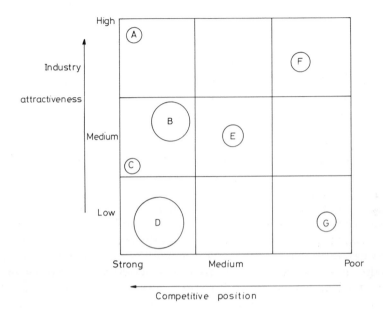

Fig. 1.5 *Portfolio analysis: a 3 × 3 matrix.*

1.5 Competitive strategy

A criticism could be made of the techniques described so far in that they do not consider sufficiently explicitly the position of a company's competitors and the actions they may take in the future. A company is seeking not only to achieve harmony with its environment, but also to succeed in that environment in competition with other organisations. In theory one could visualise two equal competitive companies which, given equal resources, could be expected to share the total market equally between them. But this will never be the case in reality, since individual companies will have different strengths and weaknesses over the whole range of these activities. Porter, in his classic book *Competitive Strategy* (1980), lists ten areas to be considered. These are:

Finance and control
Product line
Target markets
Marketing

Sales
Distribution
Manufacturing
Labour
Purchasing
Research and development.

It will be noted that five of these areas – product line, target markets, marketing, sales and distribution – are market related. This is not surprising since it is in the market place that competition for customers occurs. Without customers none of the other activities of a company's operations, however efficient, is of any value.

It is useful here to introduce the concept of *competitive advantage*. One can consider an analogy with football where the aim is to win rather than to draw or lose the match. It is of course a team effort and all players have an important role to play. But can we identify the key to success? It must be the ability to score goals since that is the only way by which the match can be won. The goalkeepers and the backs can only ensure that the team does not lose. So it is with business. It is essential to identify which of the factors listed by Porter is crucial for achieving success in a particular situation, for it can be any one of them. Let us consider finance as an example. If there is an adequate financial resource available for all the competing companies, the additional financial strength of one of them is largely irrelevant. However, in another situation the supply of investment funds available to one company may enable it to achieve results out of the reach of its competitors. Thus the aim is to draw a relief map, identify where one has a key advantage and exploit it in order to compete where others do not have the means. This is sometimes referred to as a *differential* advantage.

In contrast to the competitive advantage one must realise that there are also competitive weaknesses which must also be analysed fully. This analysis has a great deal in common with the conventional strength and weakness analysis described earlier. It is the difference in emphasis which is important in that it focuses attention on the company's position *vis-à-vis* the competition and aims to identify the one or a few elements which are critical for success.

In order to establish one's competitive position it is necessary to have an understanding of your competitor. This must be realistic and not coloured by wishful thinking. By role playing it is possible to apply all the techniques described earlier to assess his position and the strategies which logical analysis might lead him to adopt for the future. Although detailed information on the competitor will not be available, there is usually sufficient to enable a sensible analysis to be undertaken. This can provide valuable insights, but is an area of strategic analysis neglected by many companies.

As we have stressed throughout a total view of the organisation is essential. Engineering and technological capabilities are a vital input to the process but they cannot be considered in isolation. A superior technology is of no value if the resources to engineer it into products which can be manufactured, distributed and sold are unavailable. Such considerations may at times lead to novel strategic solutions. One example is the formation of a joint venture food company to exploit biotechnology between a chemical manufacturer which processes the technology and a food company which understands the consumer food market. This example illustrates another attribute of top management which has not been mentioned so far: that is, the need for creativity. Analysis only provides a framework for identifying what should not be done and for evaluating proposals for what might be done. It may indeed trigger creative thinking, but the ideas themselves have to come from people.

1.6 What business are we in?

This is a question which is often asked in the literature. The classical example quoted is that of the American railroads; the argument is that if the question had been asked the answer would have been *transportation*, suggesting diversification (possibly into air transport) as the railways declined.

It is a valid question to ask and focuses attention on the key expertise of the company in marketing, manufacture, a particular technology or finance. For example, an investigation into one highly diversified company which appeared to have no integrating thread revealed that tax and the exploitation of opportunities arising from fiscal change was the integrating factor. Exploitation of the taxation system may seem an obscure definition for a business which incorporated a number of manufacturing companies, but it did reflect the corporate reality.

This is a highly conceptual approach, but it can be useful in revealing where the key expertise lies. It may not be of much help, however, in deciding what to do. For on closer examination few of the skills required to run a railway might be those essential for the operation of an airline. It may assist in widening the boundaries of corporate vision, but the problems of implementing corporate change cannot be lightly dismissed, however attractive other opportunities may appear conceptually.

1.7 Response to outside pressures

A more behavioural view of the role of top management considers the pressures put on them from a variety of sources. After all it is people with whom management spend most of their time. They all want something and must be kept reasonably happy if their co-operation with the organisation is to be maintained. Some of the major groups whose co-operation the management must ensure are: shareholders, bankers, government, suppliers, workers and customers (Fig. 1.6). Examination of the outer ring of Fig. 1.6 suggests that all these pressures can be represented in financial terms. Thus the role of management can be likened to the exercise of judgement in a range of trade-offs. Although there are non-financial incentives and other less quantifiable means of achieving co-operation, there are few of them which in the final analysis do not involve the allocation of financial resources.

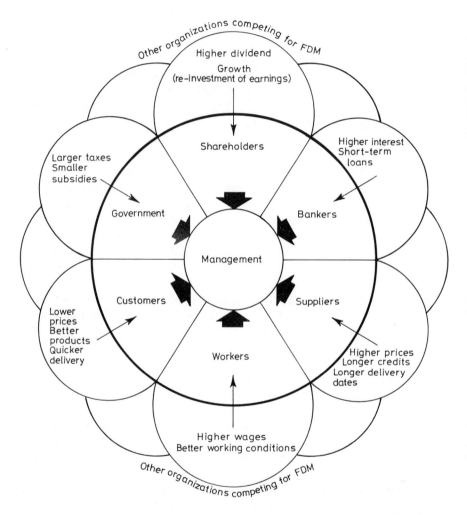

Fig. 1.6 *Pressures exercised on management by the factors of decision-making. Source: Twiss and Weinshall, 1980.*

The process in Fig. 1.6 can be likened to a zero-sum financial exercise where all the cash flows into the company are balanced by flow out of it. This is true if retained earnings are regarded as shareholders' funds earmarked for investment for future growth or to be held in financial investments to reduce the risk to their initial investment in the company. Thus management's role is seen as one of deciding the proportions in which the money available is allocated to the various parties with a call on them. It is an elaborate balancing act. If one group is given too little then co-operation is lost. This may occur because another group has been awarded more than the minimum which it would have accepted in order to remain co-operative.

This can also be likened to a power game where the opposing forces are not equally strong. Thus management might be considered wise to favour those who can be most demanding because of their willingness to use their current strength. This position will change continuously. At one time there may be a sellers' market in which the customers are weak and the suppliers strong; in such circumstances less attention need be paid to customer service since the customer has little choice. At other times the employees and their trade unions may be strong, and then loss of co-operation could have a greater impact on the organisation than at other times, such as in a recession when the loss of a few weeks' output through a strike may indeed be advantageous to the company.

It may appear that this is a cynical approach, but it is not really so. It is, however, realistic and the objective should be regarded not as 'doing down' a particular group but as keeping all groups satisfied. Undue exploitation of strength *vis-à-vis* any one group is not in the long-term interest of the organisation since it is likely to increase their unwillingness to co-operate at a later date when their relative position is stronger. For the company is dealing with people, who have long memories.

One can also see that the aim is not to maximise the return to the shareholders – a view which was often held in the past. It recognises that there are a number of stakeholders in the business, all of whom must receive an adequate reward for the contribution they make to it.

1.8 Is a strategic approach necessary?

A number of techniques for directing a business have been described. It is now necessary to ask whether this is necessary, for there are a number of successful companies where the degree of formality implied by these techniques does not appear to be practised.

For example, it is exceptional for small entrepreneurial companies to engage in any formal strategic planning. There are two main reasons why this should be so. Firstly, the degree of complexity is low. The range of products is likely to be small, as are the number of customers. The entrepreneur is intimately involved in the day-to-day business and can be expected to have a good understanding of the whole of the limited environment in which he operates. Secondly, he has a thorough knowledge of the totality of the organisation within which he has developed. These two factors combine to ensure that a high proportion of the knowledge needed to run the business is concentrated in one head, with the consequence that there is no great need to establish systems for the collection, communication and integration of information. The failure rate of small entrepreneurial companies is high, but this is usually the result of a lack of management skills, particularly in marketing or financial control, rather than the absence of strategic management.

As a company grows its operating environment will become larger. There will be a need to establish separate departments for manufacture, marketing and finance, leading to a delegation of duties. No longer will it be possible for one man to possess all the knowledge required to run the business. Formal systems are needed to deal with the increased complexity; hence arises the stimulus for the introduction of a variety of formal managerial systems, of which strategic management is but one. A further stimulus comes with the eventual retirement or death of the founding entrepreneur, in some cases at a time when the company is already large. This removes the one man who understood the whole business and had an intuitive feel for its development through time. It is extremely difficult if not impossible for one man to replace him. The new chief executive must rely much more upon others, and systematic approaches are required to bridge the gaps and provide a common approach to the analysis and use of information.

At a later stage in the growth of the business it faces other dangers. The level of complexity increases further and in many companies the activities cover a range of diverse businesses. The strategic management or corporate planning function can grow large, rigid and remote from the actual operations. It is in such cases that the publicised failures of strategic planning have often occurred. The successful companies have approached this problem by loosening their rigid centralised controls and pushing much of the decision-making to lower levels of the organisation. This was noted in the companies analysed in the book *In Search of Excellence*. One also notes the trend in recent years for European companies to disband their large head offices and replace them by small groups of people concerned with a few key parameters. There are a variety of ways in which this can be done. GEC, for example, concentrates on monitoring a few key financial ratios from decentralised profit centres. Shell, on the other hand, uses scenario planning as the main element of its strategic management, with a hierarchy of scenarios running down throughout the business.

Summarising the above, we have identified three stages in a company's growth characterised by different forms of strategic management:

Entrepreneurial Usually but not necessarily small, with decision-making centralised in one person and an absence of formal strategic analysis.
Centralised Normally medium to large firms with control centralised in the head office, often supported by a large central planning activity.
Decentralised Large to very large and often multinational, with a smallish head office concentrating on major strategic issues and more detailed strategic planning at divisional level.

These categories are, of course, generalisations and many other forms of organisation will be found. Although there will always be some exceptions, these stages of growth with the appropriate strategic approach are likely to characterise the successful business as it develops. Many failures have occurred when businesses have not made the necessary changes in their organisation necessitated by their development.

The strategic management techniques must of necessity be based upon an analysis of the information available at the time. They cannot take account of changes in the outside world that could not have been foreseen. These changes may come from both the wider business environment, such as wars, exchange rate fluctuations and raw material prices, or in the immediate corporate environment, such as the bankruptcy of a competitor or the loss of a major contract. Short-term opportunities must be seized when they offer themselves, provided they make strategic sense; inevitably they will necessitate an amendment to the existing plans. Indeed some successful companies would appear to depend almost entirely upon their opportunistic exploitation of such circumstances, relying upon the entrepreneurial flair of their chairman or chief executive. Our concern with intellectual approaches to business management should not blind us to the benefits that can occasionally be gained from such decisions. Earlier, uncertainty was equated with risk, but the other side of the coin must not be ignored. Flair might be regarded as the intuitive grasp of strategic essentials by one man without the use of formal aids. Nor should we forget that the word 'enterprise' is often rightly used to describe a business organisation.

The aim must be the marriage of enterprise and intellectualisation. Enterprise without strategic analysis can lead to dramatic failure. The use of strategic analysis without enterprise can bring about paralysis and a lingering corporate death.

1.9 The role of the engineer

Engineering is only one of the resources which a company harnesses in order to achieve its corporate objectives. It can only justify the investment made in it by the contribution it makes. A dynamic company should have this contribution under constant review, for there should be no guarantee of the continued existence of the engineering function in R & D, design or manufacture. There are many companies which at one time engaged in all three of these technical activities but which have since withdrawn from one or more of them. There are alternatives to in-house activity. R & D can be replaced by contract R & D. Product design can be supplanted by the manufacture under licence of another company's products. Even the products themselves can be purchased from a supplier and resold, sometimes after relabelling. The overriding consideration must be whether the investment in engineering yields a greater benefit than the same investment in another activity – a takeover, or marketing. Fortunately for many engineers an evaluation in these terms is often neglected and is only undertaken when the company is in financial trouble.

Engineering management has a contribution to make at three levels. It must ensure that the potential of engineering knowledge is fully exploited at the corporate level. As a result resources will be allocated to it in order to further the corporate objectives where they have a technical dimension; that has been the theme of this chapter. Once these resources have been acquired they must be allocated between competing demands for them, since it can be assumed that there will be more potential engineering uses than the funds available. This leads into the topic of Chapter 2 which is concerned with the formulation of a technical strategy. Finally it must be ensured that the individual projects within the technical strategy are managed in a cost-effective manner.

The first of these roles must be played within the overall strategic framework. In many companies which depend heavily upon their engineering expertise the strategic application of this expertise has not been exercised to the extent that it should have been. Engineers often believe that this is due to a lack of understanding of technology by boards of directors dominated by non-technologists. On the other hand these other directors criticise the narrow, specialist, non-business orientation of the engineers. There is no doubt an element of truth in both these arguments; this book is aimed at addressing the second of these complaints. A criticism of some other companies where the boards of directors are dominated by engineers is that the broader business context is insufficiently understood. In both cases there is a clear need for engineers who are broadly educated in business.

In summary the main points emerging from what has been discussed so far are:

The complexity of business decisions Judgement has to be exercised over a wide range of factors, many of which cannot be quantified.

The contribution of strategic analysis A number of techniques have been described. They are not alternatives; each is valid and can help to give insights which aid the decision-making process. They are not ends in themselves, but help to structure the information available and assist in the communication necessary for the meaningful discussion of strategic alternatives.

The changing business environment Management emphasis must adapt to the changing business and competitive environment. Thus the policies which are appropriate for today's circumstances are not necessarily valid for the future. Management must be proactive. In particular, technology is becoming of increasing importance in most businesses.

Risk and uncertainty The businessman cannot avoid risk but he must analyse the uncertainties and their associated risks so far as is possible. There will always be a residual risk. Flexibility must be built into strategies wherever possible.

The evolution of the company As the business progresses from an entrepreneurial to a centralised and to a decentralised divisional organisation, so does the degree and type of strategic management need to adapt.

Successful businesses result from people A number of points have been covered here. The corporate culture is important in that it imposes constraints on what is acceptable within the organisation. Although structured intellectualisation of the business is important it has no meaning in the absence of action and commitment to the achievement of objectives. Creativity is necessary for the generation of novel alternatives. Flair and enterprise are important ingredients of success.

Management is concerned with change Although this is implicit in the previous points it is of such overriding importance that it must be stressed again. The job of every manager is to change his organisation and to lead it into the future. Good engineering is meaningless unless it is applied to produce profit in the conditions of the future.

The business is an integrated whole The interactions of the various parts of the business are so intimate that they cannot be separated. The whole is greater than its constituent parts, all of which must be understood by all managers irrespective of their initial training and specialisation. Important as this was in the past it is becoming increasingly so under the all-pervasive impact of microelectronics.

Chapter 2
Strategy for engineering and technology

2.1 The need for a strategy

From our consideration of strategy at the corporate level it was apparent that it serves three main functions:

1. To ensure that longer-term trends are reflected in managerial decisions
2. To integrate major decisions in relation to the needs of the organisation as a whole
3. To identify areas where investment must be made to effect major changes.

These needs are equally relevant in technical decision-making. Since technology is the major stimulus for change in many organisations, a technical strategy analogous to the corporate strategy is essential.

A major weakness in many engineering organisations is that in the absence of a long-term strategy decisions are made on a day-to-day *ad hoc* basis leading to incremental changes. These will often give short-term improvements but do not enable technology to makes its rightful contribution to the achievement of the corporate objectives. It is easy to see how this can occur. The technical resources of a company are represented by its current equipment and manpower resulting from decisions made in the past. For the most part these are committed to ongoing activities which cannot be easily abandoned. By their nature they are difficult to change. In R & D the main resource lies in the skills and knowledge of its qualified technologists. If it becomes necessary to change the emphasis from, for example, mechanical or electrical engineering to electronics, or from chemical engineering to biotechnology, the lead times can be extremely lengthy. In manufacture, by contrast, the major investment lies in the capital equipment, much of which will have a lengthy remaining working life. Thus it can be seen that a high proportion of an engineering budget is unavailable for introducing changes, and that a distant planning horizon is essential in order that the limited resources that can be spared for new activities are spent wisely.

Although the strategy is the result of an intellectual analysis it is only implemented through managerial decisions and actions. In practice this takes the form of specific development projects or the purchase of items of equipment. It can also be expected that in most organisations the total engineering budget is relatively inflexible. As a consequence the strategy is concerned with the allocation of those resources between product or process development and manufacturing machinery and between the technical personnel required for current and future activities. Thus all projects, whether involved with present or with future needs, are competing for a share of this budget. A choice has to be made, and it is the strategy which should determine this choice.

This chapter discusses the factors to be taken into account when formulating the technical strategy and shows how they can be combined in a systematic series of approaches. The concepts are almost identical with those described in Chapter 1. Nevertheless there are a large number of considerations which need to be weighed and analysed in detail. The technical strategy is not, of course, created in isolation; its purpose must be to further the corporate strategy which is the most important input to the technical strategy. Nevertheless it must be stressed that strategic thinking at the technical level is not solely a matter of allocating resources to achieve corporate objectives. It must not be passive. Corporate opportunities based upon the exploitation of technical potential must be identified and assessed. Thus to be effective the two strategies must be interwoven as part of an iterative process.

A strategy is not created in isolation. Not only have certain resources been pre-empted by past decisions. There will also be an inherited strategy, which can be either stated explicitly or deduced from a study of the organisation's history. To a large extent these strategies will form part of the accepted wisdom of the senior engineering managers. For a technical strategy to be of value these traditional values must be challenged. If this is not done the strategy can be counter-productive in that it will provide a formal basis for justifying a continuance of past policies. The term 'strategy formulation', although commonly used, can be misleading since it might be interpreted as a once-for-all analysis carried out without reference to the existing dynamics of the organisation. It is preferable to think in terms of 'introducing strategic change'. In order to discharge this role it is essential that the technical departments allocate some of their funds to exploratory work. This may be related to the identification of new technologies and the initiation of some feasibility studies to assess whether they do present a strategic opportunity. Technical judgement alone can decide which areas should be investigated. The measure of a good technical director is his ability to find these opportunities and, once they are found, to present them in a convincing manner to the top management.

This process is illustrated diagrammatically in Fig. 2.1. It will be noted that although information flows in both directions the financial flows are from the top downwards. In the figure three strategies – technical, R & D and manufacturing – have been shown. This will rarely be the case in practice. If all technical activities are combined under a technical director there will be one technical strategy covering both R & D and manufacturing. However, if R & D and manufacturing are separated at board level they will have separate strategies. Because of the integrating nature of the information technologies, R & D and manufacturing are being drawn closer together and an increasing number of companies are now combining the two functions under one technical director. In the past fewer companies have had an explicit manufacturing strategy than an R & D strategy.

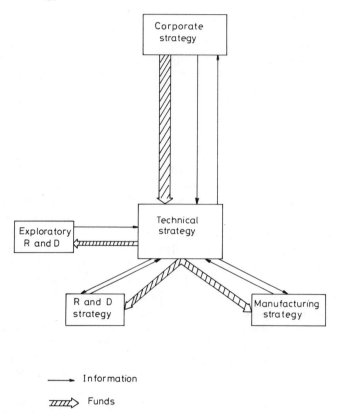

Fig. 2.1 *Technical and corporate strategies.*

2.2 Product or manufacturing investment?

Where both R & D and manufacturing are combined within one technical function, a consideration is forced of the relative allocation of funds between the two activities. This is an important strategic consideration which is often neglected. The fact that a company has been engaged in both R & D and manufacture in the past should not be taken as a sufficient reason for this to be carried into the future. It is extremely difficult to take a decision to eliminate or severely curtail one or both of them, if for no other reason than that there are strong vested interests involved.

Table 2.1 *Product development, manufacturing and marketing choices*

Strategic importance	Activity		
	Product development	Manufacture	Marketing
High	Design and develop full range of products	Manufacture all products and components	Market whole range to the end consumer
Medium	Some design Some licensing Joint ventures	Assembly only Some purchase of end products	Market to industrial companies
Low	No product design	No manufacture	Subcontractor to one or a few customers

Note All combinations are feasible but the strategic emphasis is likely to be high where a competitive advantage exists.

Table 2.1 lists some of the most important alternatives in the areas of product development, manufacture and marketing. Most traditional engineering companies are involved in all three activities. The reasons for this may be largely historical, dating back to the time when the strategic options were limited. Today there is a greater choice available and the desirability of being engaged in all three activities must be reviewed critically. The ultimate justification must be whether investment in, for example, manufacturing will yield an equal or greater return than will be gained by the purchase of the products and investing the funds released elsewhere in the business. In practice a decision to abandon an existing function is rarely made and would not occur without considering a wide range of factors, some of them hard to quantify, within a long-term perspective.

A number of the considerations explored in Chapter 1 must be taken into account when weighing where the main strategic thrust should be made. Amongst these it is important to assess where the company's competitive advantage lies. For example, if the company's main strengths lie in marketing it may be desirable to invest more heavily in this area, but almost inevitably at the expense of other activities. This might entail the purchase of some or all of the products from other manufacturers; this policy has been adopted by domestic appliance firms, which have exploited the low manufacturing cost of foreign competitors by purchasing their supplies and attaching their own labels to them. In computers, Amstrad has not attempted to compete directly with its fully integrated competitors, many of whom have experienced serious financial difficulties; its success has been built upon product development, good marketing and the subcontracting of manufacture. Rover cars, faced with the need to make large investments in both product development and manufacture, opted for the latter with the focus on assembly. Thus it co-operated with Honda on product development and increased its purchase of subassemblies such as gearboxes.

These examples illustrate only a few of the alternatives available but do underline that choices can be made. (Fig. 2.2). The aim is to emphasise those areas where the distinctive strengths of the company give it a competitive advantage, building on strength rather than supporting weakness. This has clear implications for engineers, be they in product development or manufacture. The continued existence of an engineering activity is only justifiable in relation to the contribution it makes to meeting corporate objectives compared with other areas of the business.

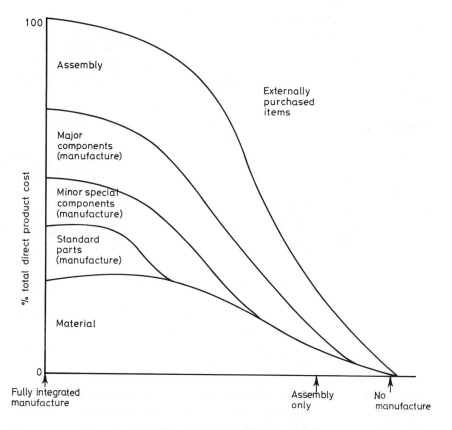

Fig. 2.2 *Involvement in manufacture: a strategic choice.*

Technology and engineering can be employed both in the development of new products and processes and in manufacturing facilities. These alternatives may not be commonly regarded as competitive if the budgets are separately negotiated. Nevertheless they are competing for funds from the same source, and decisions must be made in the final analysis from an assessment of the benefits to be obtained.

2.3 Horizontal and vertical integration

Horizontal integration is the term used to describe the expansion of a company's market share by organic growth or the acquisition of competitors. Vertical integration is of more importance to the engineer. This is the extent to which the company should be engaged in the total spectrum of activities from marketing to the final consumer to the manufacture of all the inputs to the product. Some of the considerations have already been considered in relation to the discussion of Table 2.1. There are, however, two other factors which must be taken into account:

Vulnerability Where a company is purchasing some of its components from a fully integrated competitor it is vulnerable to that company's pricing policies if there are few alternative suppliers. For example, a paint company purchased its raw materials from a fully integrated supplier which priced its supplies to make its profit from the raw materials rather than the value added in converting them. This ensured that the profit margins of the non-integrated firm were low. In such cases there is a strong argument in favour of integration.

Innovation An investment in a manufacturing operation discourages the introduction of a new technology which would make it obsolescent. This effect can be seen by comparing the American and European motor industries. The American industry is vertically integrated and as a consequence has introduced few technical innovations. Ownership of a carburettor manufacturing factory does not encourage the adoption of fuel injection. By contrast the European industry is not integrated and can purchase its components from a variety of suppliers. If one of these develops an innovative new product then it can be adopted by the car assembler without any financial penalty. This stimulates competition between suppliers and encourages innovation.

No generalisation can be made regarding the desirability of vertical integration since it depends upon the characteristics of the industry. Nevertheless it is a factor which cannot be ignored at both the corporate and technical strategic level.

2.4 The industry/technology life cycle

A technology-based product area, or an industry concerned with one main product area, experiences a life cycle (Fig. 2.3). In the early days of the technology (phase A) the potential of the technology has been exploited to only a small extent. There is a rapid succession of new products each with a considerably improved technical performance compared with its predecessor. The competitive advantage lies with the product incorporating 'state of the art' knowledge; success is highly correlated with R & D strength. The performance jump between products enables the innovating company to achieve a high market share and the rapid recovery of its development costs through premium pricing. Because the technology is still advancing rapidly the initially successful company is in a vulnerable position since the advantage gained is only temporary. The company aims to maintain its lead by also introducing the next generation of product. But this may be difficult to achieve in the state of technological

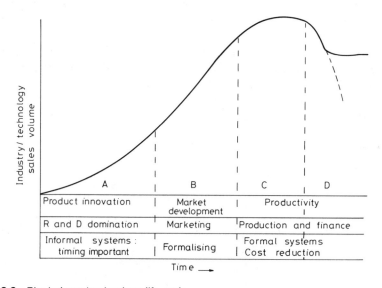

Fig. 2.3 *The industry/technology life cycle.*

anarchy typical of this phase, where many firms, often small and entrepreneurial, are exploring radical applications of the new technology.

It will be seen that the keys to success in this phase are technological excellence and the ability to develop new products quickly. The performance advantages of the new product are what count in the market place, where the customers are more concerned with this aspect than with price and reliability. Of course, both price and quality must meet acceptable standards but they are not the main criteria in making the purchase decision. Since the performance enables a premium price to be charged, development and manufacturing costs are of secondary importance. If undue attention to cost control delays the launch of the product it can be a misguided form of economy. In order to succeed the strategy must be heavily weighted towards R & D, as a consequence of which the R & D director will have a high relative status since it is his contribution which provides the key to success.

As the technology advances into phase B the degree of turbulence diminishes and the pace of performance improvement slackens. It is at this stage that one or a few dominant designs establish themselves. These reflect the inherent advantage of a particular technical solution which was not apparent in the earlier stages of development; the other approaches fade from the scene, for example the steam car or the flying boat. The requirements of the customer are now more easily understood and they can become more selective in their choice from the range of not dissimilar products offered by competing suppliers. Thus the contribution of the marketing department increases and market considerations will have a more important strategic influence than previously. Some of the growth will be entirely market led such as expansion into new geographical areas. Although new products continue to appear they will be less radical departures than in phase A.

The most significant engineering impact will come from the identification of market segments. These are groups of customers with tastes similar to each other's but different from those of other segments. Product variety increases, and the role of the technical departments will be the design of a range of products to meet all or a few of the selected segments. In many cases the product specification will result from detailed market research. It is, of course, vital that the market is considered throughout, but in phase A it is more likely that the technologist will appreciate the market opportunities, whereas in phase B it will be the marketing man. Development is now market driven rather than technologically led, with a consequent rise in the status of the marketing department and its director.

The next stage, phase C, is reached as the technology approaches maturity. Most of the potential of the technology has been fully exploited and considerable R & D investment is needed to obtain marginal performance improvements. Thus it becomes exceedingly difficult to improve upon current product designs and the competitive offerings are hard to distinguish. The characteristics of the manufactured product closely resemble those of a commodity. Heavy advertising expenditure may be devoted to establish a brand image to differentiate the product from that of the competitor, but in technical terms the products are virtually identical. The customer is now highly sensitive to price and quality. This will have an important impact on the technical strategy which must be focused on manufacturing cost reduction and improved quality. Manufacturing investment to meet these objectives will take predecence over other technical considerations. R & D will make its contribution more from process designs and improvement than from new product development. In terms of organisational status the finance and production directors will rank high since it is their contributions which are vital for success. Control systems will be formal and inflexible.

Many of the traditional engineering companies have reached phase C, and it is in companies of this type that a high proportion of a nation's engineers are employed. It is not, however, a plateau on which many companies can rest for long. There are a number of alternatives as a firm moves into phase D, as follows.

Industry stability: After a shake-out of companies a few or even one monopoly supplier remains. Usually margins are slim and competition is fierce. Normally, as with steel, a long-term decline ensues.

The industry disappears: Innovation by invasion has been mentioned earlier. Although the market need remains it is satisfied by an entirely new form of product which substitutes for the old, e.g. cut-throat razor, wooden clogs, slide rule. When faced with this situation the main strategic choices are to diversify into a new business based upon financial or marketing expertise, to satisfy the existing market by adopting the new technology, or to develop new markets based upon the company's traditional technology.

A re-examination of the feasibility of using technology to enhance performance: Under the threat of decline a fresh incentive is given to the search for improved performance by the application of technologies developed since the design of the original product stabilised. A common response is to harness technology to increasing the in-use life of the product, e.g. stainless steel razor blades, long-life batteries, corrosion resistant cars, and radial ply tyres. This increase in product longevity enables an unchanged level of consumer demand to be satisfied by a lower industrial output, owing to a significant decrease in the size of the replacement market upon which the industry is heavily reliant once it has reached maturity. Thus although the usage of the product is largely unaffected the industry moves to a lower output level at which it may stabilize again. This use of technology is perhaps inevitable but it poses a considerable threat to the companies engaged in it. As the total volume falls some companies must withdraw either voluntarily or by default. Unfortunately human optimism is such that each company believes that it will be the survivor.

The business environment changes, giving a new impetus to innovation: Most products possess a number of attributes desired by the customer. For example, with a motor car there may be speed, comfort, safety, durability or economy. Each individual has his personal weightings for these attributes and the aim of market segmentation is to identify a sufficient number of potential consumers with similar tastes. In a stable environment these segments are unlikely to change to any significant extent. However, when the environment changes this stability is unbalanced. Thus when fuel costs rose rapidly a new stimulus was given to fuel economy. In the motor industry this encouraged the development of electronic engine management systems and the increased use of light materials. It might be argued that these developments would have occurred in any case, but it is doubtful whether they would have been introduced at the same rate without the increased emphasis on fuel economy.

The stimulus can come from any part of the business environment – economic (e.g. tax changes), social (e.g. environmental concern) and political (e.g. safety legislation) as well as new technology. The technical strategist should be scanning and noting such developments and examining how technology can be utilised to turn what at first sight might appear to be threats into opportunities.

The exploration of technological limits: In phases A and B the scope for innovation is reduced as the potential of the technologies embodied in the product becomes fully exploited. The performance ceiling is approached. Thus the performance of the piston aircraft engine was limited by the aerodynamic characteristics of propellers. The thermodynamic efficiency of the internal combustion engine is a function of the combustion temperature, restricted by the properties of the engine materials. These limits to performance for a particular design configuration are a consequence of the physical properties of part of the system (see Chapter 3). If, however, these constraints can be removed through technological advance a new performance growth curve becomes feasible. Thus in the case of the aero engine the development of high-temperature steels for turbine blades provided the capability necessary for the jet engine to compete with the piston engine. Similarly the development of a ceramic internal combustion engine will lead to a period of further growth in its thermodynamic efficiency which for many years has been on a plateau.

It is useful to distinguish between this type of development and 'innovation by invasion'. In this case there is no major threat to the skill base of the company so that the innovation is within the competence of existing companies, although there could well be a major impact upon the relative competitiveness of individual companies within the industry.

Consideration of the life cycle and the changing nature of the technological contribution is of great assistance in formulating the technical strategy. In summary it can be seen that:

1. The emphasis of technology moves progressively from radical new product development through product improvement to process development and manufacturing investment. It must be stressed that this is a matter of emphasis, since all aspects are relevant at all times, for example process development and manufacturing investment are required during all phases but they are relatively less critical for success in phase A.
2. An examination of the life cycle is a valuable indication of when a shift in the technical strategy becomes desirable.
3. The most difficult and complex strategic technical alternatives occur in phases C and D, which is typical of the situation for many large engineering companies.
4. Each phase is associated with a different management style – from entrepreneurial in phase A to formal in phase C – and a change in the status of senior management, often not recognised, reflecting their relative importance in achieving the corporate objectives. In terms of management there is no *right*

style, only one that is *appropriate*; furthermore it changes with time. The four phases have sometimes been characterised as: enthusiasm, contentment, complacency and dismay.

5 Although the choice of the strategy is important it is of little value unless it leads to the successful conclusion of engineering projects, which depends upon the adoption of appropriate management styles and systems.

The industry/technological life cycle as described above provides a simple model for those industries based on one or a limited range of similar products. Thus it is easily applicable to industries such as base metals, motor car manufacture and much of heavy engineering. The major problems arise when it is desired to initiate new technological activities when the major products approach maturity. For while there may be an intellectual recognition of the need to divert more resources into new technology and product development, the achievement of this aim may be frustrated by the rigid control systems which have been installed as the company has evolved. This topic will be discussed in more detail later, but it can be seen that this can explain why many technological innovations are introduced by the smaller companies rather than the industry leaders from whom such developments might be expected.

This problem becomes more acute with the multiproduct company, particularly if its products are based upon different technologies. In such cases the overall technical strategy must reflect the requirements of a portfolio of needs with some products in each phase of the life cycle. It then becomes exceedingly difficult in practice to strike the right balance between management styles. As mentioned earlier it is the new product development area which usually suffers through the imposition of company control systems developed for the rigorous cost control of mature products. A few companies (e.g. 3M, Texas Instruments) have been notably successful in achieving an appropriate balance, but these are the exceptions. These problems extend to manufacture. In the mature market product change is infrequent and uncertainty is low; the production workers are proficient in operations which have changed little in years. With the introduction of radically new products new skills must be acquired, but more importantly the workers must be capable of accepting frequent modifications to specifications and working practices. This produces stress and can cause industrial unrest, particularly if the payment systems are related to the conditions of uninterrupted, unchanging work practices.

How should these difficulties be overcome? They present a challenge to management. Although some successful companies have found a solution to the management of a mixture of products at different phases of the life cycle under one roof, they are the exception. Serious consideration must be given to the desirability of introducing radical new products in new facilities geographically separated from existing production. In this way can be achieved the flexibility of the small innovative firm, supported by the financial resources of its large parent.

Strategy for engineering and technology

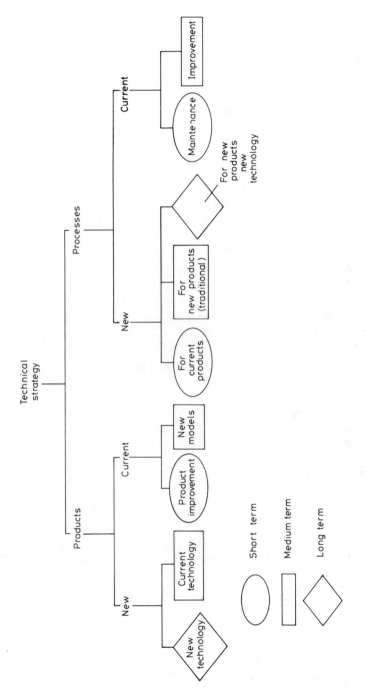

Fig. 2.4 *Technical choices.*

38 Strategy for engineering and technology

This discussion of the life cycle has digressed from what may at first sight be regarded as technical strategy formulation. It must be recognised, however, that strategy is not an analytical activity divorced from the company's manufacturing operations. It affects company organisation structure, managerial styles, the siting of facilities and employee relations. What may appear to be an excellent strategy will be frustrated in practice unless all the organisational implications are thought through and reflected in an integrated comprehensive plan. The engineer cannot confine his attention to technical issues alone.

2.5 Developing a technical strategy

Having examined some of the broader issues surrounding the role of technology in an organisation, it is now possible to explore how these can be incorporated within a more formal system for the developoment of a technical strategy. We have seen that the strategy results in the allocation of resources to areas of activity within which it is implemented by specific projects. In this and the following sections the main categories of technical expenditure will be discussed and the methods for allocation described. In later chapters the choice of projects for new product development and manufacturing investment will be explored.

The main categories of technical investment are shown diagrammatically in Fig. 2.4. It is seen that they can be broadly divided into the short, the medium and the long term, although there will inevitably be some degree of overlap.

There is always an urgency associated with the solution of short-term problems. Plant maintenance, for example, is essential and there can be little argument about the resources devoted to it, particularly in respect to routine preventive maintenance. A proportion of effort will be devoted to rectifying breakdowns. But why do the breakdowns occur? It could be because too little effort has been devoted to routine maintenance in the past, or because of weaknesses in the plant design which could be overcome by engineering improvements, or because the plant is approaching the end of its useful life and should be replaced by a more modern design. Thus even in this rather mundane example it can be seen that there are alternatives covering different time spans. It might be thought that a cost–benefit analysis would lead to the optimum solution. In practice it is not so easy for a variety of reasons. The decision is clouded by the normal division of budgets into operational (or revenue) items and capital expenditure. In terms of the cost–benefit analysis this distinction may appear irrelevant, but it is not so in relation to cash flow. The shorter-term solutions involve relatively small annual expenditures on a continuing basis, whereas the long-term investments incur a major cost with a correspondingly large impact on cash flow. The limited availability of cash implies that the adoption of what might appear the optimum solution absorbs money which would otherwise be allocated to some other project. Of course, more money in total might be made available but this might require the company to acquire the money by raising more share capital or by debt. This simple example draws attention to four elements of the problem:

1. Every technical decision can be viewed as having short-, medium- and long-term alternatives.
2. An engineering expenditure in one area usually involves a sacrifice in another.
3. An engineering decision involves both profitability and cash flow implications.
4. The allocation of additional funds relates directly to the company's financial strategy.

The imperatives of the short-term problems can easily detract from the long-term needs. For the individual manager the short term imposes immediate pressures from colleagues for satisfaction. Human inertia makes it easier to carry on with what has been done in the past rather than search for more radical solutions. Furthermore, since the long term usually requires a capital investment of some sort there will usually be opposition. It should not be assumed that priority should always be given to the long term, but its importance needs to be stressed because of the strength of the forces encouraging its neglect in favour of the immediate. Thus a main purpose of strategic thinking is to direct attention to the balance of needs across a longer time span and to expand consciously the time horizons of managerial analysis.

The development of the technical strategy normally evolves with time. It is a continuous process, and only rarely will a major strategic review indicate the need for a significant reorientation of the company's engineering effort. Where, for example, one is concerned with developing the firm's technical potential over a 20 year period, it would be surprising – and indeed of little value for decision-making – if there were significant amendments at annual intervals. New knowledge which has an influence on the strategy is being acquired all the time, so the underlying forces shaping the future are changing continuously. It is not, however, practical to amend the strategy too often and an annual review is normally regarded as sufficiently frequent. It is important, however, that the review should reassess all the information to ensure that it provides the basis for a genuine reappraisal. The danger of taking the previous year's work and making minor modifications without a serious attempt to evaluate new factors must be avoided.

2.6 Status review

An organisation has reached its present state through the aggregate of all the decisions taken in the past. Thus a technical strategic review should start with an analysis of how its resources have been allocated previously, under a number of broad headings. The first time this is done some difficulty may be experienced if the data has not been analysed in this way before. This is not a serious problem since a high degree of accuracy is not necessary and judgement will often be needed. Where possible this exercise should go back several years since the dynamics of change are as important, if not more so, than a realisation of where the company currently stands. It may reveal for example that the focus of R & D expenditure has shifted from new products to new processes. There could be several reasons for this. It could have been a conscious decision, taken to meet corporate objectives for cost reduction or the introduction of new products developed earlier. It might mark a matching of R & D policy to the evolution of the life cycle. On the other hand it may be a reflection of influences of a more subtle character. In times of high interest rates, the application of financial analyses, particularly if DCF is used, will favour projects with a shorter payback period; these tend to be new processes rather than new products and small projects rather than large. This could happen solely as a result of applying the financial criteria without an explicit examination of their wider effects. Alternatively the change might have been a consequence of changing power structures within the firm, say marketing versus manufacture.

The important value of this analysis is not to make judgements on the decisions made in the past, but to observe what has actually been happening, identify the reasons for it, and question whether changes will be appropriate for the future. The headings under which this analysis can be carried out are:

1. The main areas of technical activity (see Fig. 2.4)
2. Short, medium and long term; the definition of these in terms of time span will depend upon the character of the industry and the technologies
3. Size of individual project – small, medium, large
4. The overall size of the technical budget.

The technical budget consists of several parts. Some, such as R & D, are relatively stable; others, such as capital investment in manufacturing, can vary considerably from year to year. This is because in many industries R & D is spread over a number of different projects and a considerable amount of discretion is left to the R & D director in the allocation of his budget between projects; those industries with one or a few major projects are exceptions since the budget may vary over time to meet the requirements of the individual project. In most industries manufacturing investment, certainly for large items, will be allocated on a project basis and may vary considerably between years. It can be assumed, however, that there will be some idea of what the total technical budget is likely to be for the coming year. Existing programmes will also have their individual budgets, from which it can be calculated that:

money available for new activities = total budget − the budgets for existing projects

In most engineering industries the margin available for new activities is likely to be a relatively small part of the total. It must be emphasised that this is only a preliminary figure, since neither the total budget nor the budgets for existing projects should be regarded as immutable. If at a later stage of the analysis a larger budget for strategic change is shown to be necessary there may be grounds for seeking an increased budget allocation; if this fails, reductions in the budget of existing projects may have to be considered.

2.7 Inputs to technical strategy formulation

In the previous section we have looked at technical strategy as a process for the allocation of a budget between choices. Although this gives valuable insights into the nature of the strategic problem it is of limited benefit if the money cannot be invested to yield a satisfactory financial return. It may have been revealed, for example, that very little is devoted to the general area of new technology. Although this recognition may be important, an increase in the budget for new technology is of no value unless promising opportunities to harness it for new products or new manufacturing processes can be achieved. Thus the strategy can be viewed as giving shape to an aggregation of individual projects. In the absence of good proposals for engineering projects the strategy becomes meaningless. It is all to easy to regard strategic analysis as a mechanistic intellectual exercise. But it cannot identify what should be done in detail, although it can suggest the directions in which attention should be focused. By providing a framework it imposes limits on what is acceptable, but to be fully effective it must motivate and spur the creativity of all engineers in directions which enable it to be fulfilled.

Thus three main flows of information into the technical strategy formulation process can be identified:

Top-down, flowing from the corporate objectives and strategy
Bottom-up, representing the ideas generated at all levels in the engineering function
Out-in, to capture the technological, environmental and competitive factors from outside the company and from internal sources such as marketing and finance.

These inputs are shown in Fig. 2.5. Each of them must be analysed individually and then in relation to the others. It is thus an interactive process similar to that followed in developing a corporate strategy. Each of these inputs will now be described briefly.

Corporate strategy
This must be a major input since the purpose of the technical strategy must be to further achievement of the corporate objectives. However, in the preliminary stages of developing the technical strategy this constraint should be interpreted

liberally, since it must be one of its aims to reveal new opportunities worthy of examination which might warrant a modification of the corporate strategy. For this reason it is highly desirable that some resources are devoted to technical exploration. The corporate strategy should indicate the extent to which the corporate strategic gap should be filled by new products.

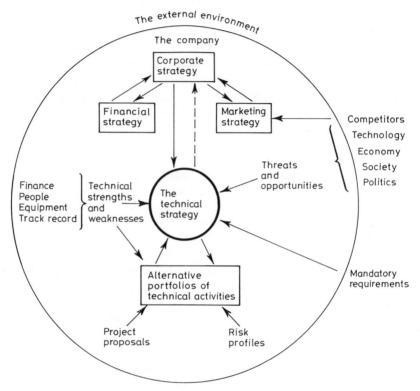

Fig. 2.5 *Inputs to technical strategy formulation.*

Marketing and financial strategies
Both marketing and finance will have their own strategies, either informal or explicit. These will be in greater detail than the marketing and financial considerations embodied in the corporate strategy. On occasions the joint exploration of strategic possibilities can lead to technical developments which otherwise might not have been feasible. The rapid diffusion of office copying technology was made possible by the decision of Xerox to rent rather than sell its machines, thereby reducing the risk to the users from technical obsolescence. Renting was also instrumental in delaying the threat from Japanese imported colour TVs into the British market, since by encouraging rental through a subsidiary of the manufacturing company a captive market was ensured. In such cases it can be seen that, by using imaginative financial or marketing strategies in conjunction

with a technological development, the drawbacks of the latter may be overcome – for example, the customer's distrust of the risks associated with new technology or lack of resources to purchase the product.

Mandatory requirements
The legal requirements of business are increasing all the time. In relation to technology, product liability and environmental controls are the most significant. They cannot be avoided. For some industries such as pharmaceuticals the cost of clinical testing is such that the pace of technological development is becoming slower and the costs inhibit the development of drugs for rare diseases where the market is small. In general this has the effect of diverting resources which the company would prefer to use elsewhere, including the exploitation of new technologies. For some industries legal and fiscal barriers may determine the location of manufacturing facilities and some of the technical services which are associated with them.

The impact of legislation also offers technical opportunities for the enterprising company. The introduction of pollution control regulations in many countries led to the development of the platinum catalyst car exhaust system by Johnson Matthey. Safety regulations also offer many opportunities for the development of new products.

Technology capture
The rapid expansion in technological knowledge throughout the world means that even the largest company is a modest player on the world stage. Thus it cannot rely entirely upon its own resources and must ensure, in so far as is possible, that it is aware of all those developments which could have an impact on its business and, where appropriate, gain access to them. This is a normal feature in the day-to-day operation of all technical departments, and in many cases the new knowledge is absorbed at the operational level in product or process design; sometimes it involves licensing.

Occasionally, however, there will be developments which may have a strategic impact. This is particularly true where the company possesses only part of the total business expertise or resources necessary to exploit the new technology. Several alternatives are possible:

Licensing This is essential where the design is covered by patent protection. It can also be useful where the company could design its own competing product, but does not have the financial resources to do so or would prefer to invest its own funds in other projects.

Contract research The use of outside organisations for development and design is increasing. This is particularly useful where the company has funds available for product development but does not wish to enlarge its own activities permanently. Or it may lack an expertise which it cannot acquire speedily and prefers to subcontract the work to an organisation that does possess it.

Pre-competitive research   In many new fields the cost of research is high. Two or more companies, sometimes direct competitors, agree to fund the research jointly and pool the results until the research can be applied to product or process development.

Shared development, design and production   In a growing number of engineering industries the cost of developing and manufacturing a new product is beyond the means of any one manufacturer. Thus in aero engines one finds a company (e.g. Rolls-Royce) in a number of consortia with its competitors on joint programmes. In aerospace the European Airbus is another example.

Many new patterns of technological co-operation between companies are emerging as the traditional structures of industry are changing under the impact of new technology. The past identification of an industry with a single technology is disappearing. For example, there are now few industries which do not require an electronics expertise. In some cases it may require a change in the relative influence of different contributors to a design. As the importance of avionics increases, and so does its proportional value in the cost of an aircraft, it may be more meaningful to consider the aircraft itself as merely a sophisticated package. This may appear an academic and conceptual argument but it can have a substantial impact on the strategic importance of the individual contributors to the end product. What is clear is that the business complexity of many engineering industries is increasing and that a much broader range of technical strategies has to be considered in order to exploit the changing technological scene and to capture those elements which are not within the total control of one company.

Technological opportunities and threats

The analysis of technological opportunities and threats is analogous to that carried out at corporate level. There is an overlap here with what has already been described above for technology capture; nevertheless it is useful to draw a distinction between the capture of new technology that to a large extent exists at present but outside the company, and developments which may occur in the future.

The most critical strategic impact occurs when a new technology fundamentally changes the nature of how a market need is satisfied. The history of industry is full of examples where a new technology has destroyed established companies within a relatively short period, e.g. electricity for gas lighting, synthetic for natural materials, the electronic calculator for the slide rule. In all three cases one notes that the new technologies were exploited by those companies which possessed the technological rather than the market expertise. This process is often referred to as innovation by invasion, since the existing market is invaded by newcomers. This presents the greatest technological threat to the established leaders in the market place since the competitive advantage on which their past prosperity has been founded ceases to be relevant. On the other hand it presents a great opportunity where a company can find a new market for products based on advances in an existing technology or a new technology; this represents a technology-based business diversification.

It might be argued that a threatened company should respond by acquiring a capability in the new technology so that it can build upon its market strengths. Although this is possible it rarely seems to happen. It was not the manufacturers of gas mantles who made electric light bulbs. In synthetic fibres it was the chemical companies such as ICI and du Pont, not the cotton or woollen companies, who exploited them. More recently there have been many examples of electronic companies invading the traditional preserves of the electromechanical and precision mechanical industries; the decline of the Swiss watch industry is a classic example.

Fortunately these invasions occur only infrequently. However, their effect can be so catastrophic that the possibility must be a central concern in any strategic analysis of technology. Of course, the opposite is also true; attention should be focused on identifying opportunities for invasion.

In many cases the invasion will not result in a simple substitution. There are several possibilities:

Substitution within the existing market In such cases, for example plastic kitchenware for metal or plastic for glass bottles, the size of the market is not changed significantly.

The market size is reduced This will occur when the product longevity increases without adding to the number of customers, as with stainless steel razor blades and socks from synthetic fibres.

The market is segmented In many cases features of the old technology will maintain their appeal to certain segments of the market, resulting in an incomplete substitution, for example the electric razor and many applications of synthetic fibres.

The market is enlarged Sometimes where the new technology results in products significantly cheaper than those it replaces there is an explosive growth in the size of the market, for example the ball point pen and the electronic calculator.

Thus there are several questions which need to be answered:

1. What new technologies could represent an opportunity or threat for the company? This implies the necessity of constantly scanning the technological scene.
2. When might they become significant? This can be a difficult question to answer. Technology forecasting (Chapter 3) can assist, but there are many examples of over-optimism – for example the superconducting generator and the video telephone.
3. What effect will it have on the size of the market?
4. Who is most likely to exploit the new technology?

Only when these questions have been assessed can consideration be given to what the company itself should do. If there is a threat it cannot be evaded. All too often failure is the result not of a shortcoming in the strategic analysis but of an unwillingness to accept it or act upon it. It was not ignorance of the electronic watch which led to the misfortunes of the Swiss watch industry but a reluctance to face up to its consequences.

Advances in a company's traditional technologies present a continuing, although less dramatic, range of threats and opportunities. This is of particular importance for a company with a limited product line, especially if development time and costs are high. Decisions for product improvement or new product development must be taken well in advance of the time when demand for the existing design falls through obsolescence. The two main alternatives are:

1 Incremental improvement of the existing design to maintain its competitiveness
2 The initiation of an entirely new design incorporating the most advanced technology.

Eventually a new design will be required. The critical decision relates to its time of introduction. The arguments in favour of incremental improvement are strong. The risks are low since the features of the design are largely unchanged. There is a smaller investment required to design the improvements and introduce manufacturing changes. The experience curve benefits of manufacture can be further exploited without the need to initiate a new learning curve. Existing customers are aware of the product and it entails a low risk for them to purchase the improved product. Nevertheless, at some time in the future a competitor will launch a product incorporating the latest technology in a new design. If technical leadership has been sacrificed, the follower may suffer a rapid fall in sales. It does not always follow, however, that technical leadership will guarantee success. There are many examples where a technically competent follower has benefited by allowing the leader to create the market with the first-generation model and then follow with an improved second-generation design. The risks inherent in technical leadership are illustrated by the brain scanner pioneered by EMI; it was quickly overtaken by American manufacturers with improved versions. A more recent example is the development of the ducted-fan aero engine which is likely to be introduced in the 1990s. What should be the response of a company with a well established product line, for example Rolls-Royce? Desirably it would wish to capitalise on its existing designs for as long as possible, but too long a delay could place it in a disadvantageous competitive position.

Since timing is so vital it must be an important consideration in the formulation of a strategy. Attention must be given to:

1 Identification of technological advances which could have a significant impact on product performance
2 Assessing the rate of progress of the performance potential against time

Strategy for engineering and technology   47

3   Monitoring the actions of competitors.

Technology forecasting can be of great assistance in identifying and assessing the progress of technology. Technological intelligence will provide information on the actions of competitors. Both must be organised systematically if the best information is to be available for decision-making.

The uncertainties of technological advance are such that no amount of analysis will yield a totally satisfactory answer. Nevertheless, the decisions cannot be avoided. There are, however, two strategies which are unlikely to succeed but which are not uncommon. The first is a total reliance on incremental development; this will lead inevitably to product obsolescence and market decline. The second involves the constant introduction of entirely new products without gaining the advantage to be derived from volume production of a stable but evolving design. The aim must be the introduction of occasional new products followed by a period of incremental advance.

Accepting the need for the interspersion of technological leaps with steady evolution leaves unanswered the strategic question of whether to adopt an offensive strategy – to be first in the market with the new technology or to be the follower. In general an offensive strategy would be more appropriate where the distinctive competence of a company lies with its technology rather than its marketing. This is unlikely to succeed in the long term, however, unless the initial advantage can be maintained through a family of improved products. All too often this initial advantage is sacrificed. The defensive strategy may be more appropriate for a company with market strength and sufficient technical strength to introduce a product that quickly incorporates the new technology and offers an improved performance. The IBM personal computer is a good example of the latter. IBM has immense marketing strength and, although highly competent technically, is not generally regarded as a leader in technical innovation.

Whichever strategy is adopted it is clear that the company should be concerned not only with what to do next but also with a long-term technological plan for a succession of new products, innovative or evolutionary. This implies that the future development potential for any radical new design must be considered in depth before initiating it.

2.8 Competitive advantage

In the discussion so far it has been assumed that the company has a clear understanding of who are its competitors. However, within an industry there will be a range of companies, most of which will be catering to the needs of one or a few market segments. One extreme is represented by the specialist manufacturer catering for the top end of the market; its product has a relatively small volume and a high price and normally yields a high profit margin. At the other end of the market is the mass producer with a less sophisticated low-priced product, relying upon high volume and low margin to generate its profits.

In terms of business competence the specialist producer is relying upon technology in product design, whereas the mass manufacturer is exploiting manufacturing and marketing strength. Most companies have a history which has identified them with one of these segments. However, because of changes occurring in the business environment the traditional stance may no longer be appropriate and should be re-examined. This has clear implications for the technical strategy. In many fields of engineering the past 20 years has been marked by an erosion of market share by European companies in the face of competition from the Far East. Initially this came from Japan with its relatively low labour costs; other countries, principally Korea and Taiwan, have followed. Japan has maintained its competitive position through a high level of productivity based upon manufacturing technology rather than labour costs. By contrast many of the most successful European companies have been those concentrating on the high value-added segment of the market. For example, in motor car manufacture BMW, Porsche and Jaguar have been notably more successful than Renault or Rover. Other companies have made a conscious shift of strategy away from the volume markets, as evidenced by ICI where an emphasis has been placed on speciality chemicals and pharmaceuticals at the expense of bulk chemicals.

It can be assumed that where the cost of unskilled or semiskilled labour is a major item in a company there is little it can do to overcome this disadvantage within its existing structure. There would be no alternative within that market segment to investment in labour saving technologies. This can only be successful, however, if it leads to a productivity level comparable with that of the best in the industry, most probably the Japanese. This is a feasible option and involves a technical strategy focused on process development and manufacturing investment.

The other alternative is to target a specialist high value-added segment of the market. This might amount to a major reorientation of the corporate strategy but could not be effected without a concentration of effort on engineering. To do this presupposes that the technological expertise exists or can be easily acquired. However, it does appear that European companies which have focused on this strategy have in general been more successful than those which have relied upon manufacturing expertise. Notwithstanding this, no ideal solution is being advocated. What is important is that the range of options is considered after a full analysis of the competitive strengths and weaknesses of the company now and in the future. The decisions based upon this analysis are highly dependent upon the contribution that can be made by engineering.

2.9 Technical capabilities

It is not sufficient to identify promising strategies, for the company may not have the necessary technical resources. In order to assess the capability a strength and weakness analysis comparable with that carried out at the corporate level must be undertaken. It should start with the capabilities in relation to the present strategy in respect of the various resources required. Some of the items are listed below, although these should be regarded as indicative rather than comprehensive:

Finance The size of the R & D and manufacturing budgets
People Numbers, skills, competence
Equipment Age and type of manufacturing plant, and laboratory equipment
Performance Track record in relation to product and process innovation and manufacturing productivity.

In order to do this analysis a detailed technical audit is required. One of the problems is that it is extremely difficult to be objective about the capabilities of one's own organisation. For this reason it may be advantageous to employ a consultant. The strategy, however, is concerned with the future where the required capabilities may be substantially different. It is necessary, therefore, to carry out a series of audits in relation to strategic alternatives. This is likely to reveal some significant weaknesses due to a lack of certain technical areas of expertise or capital equipment. These weaknesses do not invalidate a proposed strategy for they can be rectified, but often not without serious difficulty. Furthermore, some of today's strengths may be future weaknesses if they are no longer relevant.

Sometimes some simple scoring techniques are utilised in an attempt to quantify the audit. These may be useful but can also be misleading. Although competence across the board is desired, success or failure is more often equated with the presence of a distinctive strength or a serious weakness.

There are few products which do not require contributions from several engineering or technical disciplines. Normally it will be found that there is one dominant technology associated with the firm's product. It is likely that the technical director will have been trained in this discipline and that the company will have little difficulty in recruiting high-calibre engineers in that discipline. Other disciplines, critical though they are, may present a very different picture; their members may be regarded less well, their promotion prospects are likely to be poor, and as a consequence it may be difficult to recruit or retain people of the desired standard. This problem manifests itself in the product, thus in radars it is the turning gear that can give problems, in cars the electrics and in nuclear power the civil engineering.

The depth of engineering skills must also be considered. There is evidence to suggest that serious deficiencies in engineering projects leading to delays and cost overruns can be attributed to shortages of qualified engineers in detailed design. It is not sufficient to have a brilliant engineering design concept if there is insufficient depth of technical ability in the company to convert the concept into practical designs.

2.10 The risk profile

There are two main contributions to the risk of engineering activities – the technology and the market. The risk of total technical failure to achieve the specification is relatively rare. Usually the problems encountered are overcome but at a cost. Forecasting the market size and market share also present a major problem, particularly where there is no past expertise. In general the greater the departure from established products and the longer the time span, the higher the level of uncertainty and risk.

As noted in Chapter 1 risk is inseparable from business. Moreover, risk cannot easily be quantified. In spite of this it is a factor which must be taken into account when developing a strategy. Normally the technical strategy will direct funds into a portfolio of activities within which it must aim to achieve a mix of high- and low-risk projects within a combined risk which is acceptable. Furthermore, for each high-risk activity the rewards if successful must be correspondingly high.

2.11 Project proposals

Strategy, because it concerns the big issues of the business, is a prime responsibility of senior and top management. However, in engineering it is likely that the younger and less senior engineers will possess the most intimate knowledge of new technologies and their capabilities. Being creative and professional they should generate many ideas for new proejcts, some of which will have a potential which should be reflected in the strategy. For as we have noted before the strategy relies upon a portfolio of individual projects for its implementation. The strategic analysis, whilst revealing general opportunities, can only play a part in determining what should be done in detail.

This means that every engineer has a contribution to make to the strategy. In many organisations a rigid hierarchical management structure inhibits the flow of ideas from the bottom up. Where this is the case the quality of the technical strategic decision-making is likely to be reduced. However, there is a growth in participative management styles and organisational structures to encourage it. A strategy which is created in the rarified atmosphere of top management without drawing upon the intellectual resources of the whole organisation is unlikely to provide the best foundation for future success. However, it must also be recognised that many of the ideas generated at lower levels will not be acceptable for good strategic reasons.

2.12 Techniques to aid strategy formulation

It is apparent from the discussion so far that there is a bewildering number of factors to be considered when developing a technical strategy. No management techniques can by themselves provide adequate answers, although they can aid in the process by providing a systematic framework for analysis. Let us now review some of the approaches described earlier to examine how they might assist:

Gap analysis   This provides an indication of the engineering contribution in the achievement of corporate objectives over time from new products or reduced manufacturing costs. It can also be used to assess the size of the technical budget.

Technology/industry life cycle   This gives an indication of where the main technical effort should be employed – new products, product improvement or manufacturing cost reduction.

Portfolio matrix analysis   This enables the technical portfolio to be assessed. An analysis of this type closely relates to the life cycle analysis; phase A products are the stars and phase C the cash cows.

Business environment analysis   A systematic scanning of the total business environment to assess the implications for technology.

Technology forecasting   A range of techniques (Chapter 3) is available to assess the environment, to identify technical threats and opportunities and to give a measure of what technical performance may be required of products and processes in the future.

Technical audit   The audit gives an indication of the company's technical capabilities, its strengths and its weaknesses.

Competitive analysis   This evaluates the technical capabilities relative to the competitors. It can contribute to a selection of the appropriate market segment: high value-added specialist; a market niche; high volume with low margin.

Risk analysis   The uncertainties and risks of alternative strategies must be assessed in order to generate a portfolio of technical activities with an acceptable risk.

2.13 The form of the strategy

In order to carry conviction the technical strategy should be a written document containing a detailed analysis on the lines indicated above. On the other hand its only value is the actions which stem from it. All organisations will have some form of technical strategy even if it is not documented. Many technical directors make good decisions based upon their own experience without the aid of any formal procedures. Since a formal strategy document costs money it can only be justified in relation to its ability to lead to better decisions. Thus there must be constant vigilance to ensure that the process does not become over-elaborate and a justification for those decisions which would have been taken in any case.

52 Strategy for engineering and technology

The advantages of a formal strategy are that it

1. Forces an explicit analysis of all relevant factors
2. Records assumptions and provides a basis for their critical examination
3. Enables the corporate objectives and strategy to be incorporated into engineering strategies, policies and plans
4. Imposes the discipline of periodically reviewing the major areas of technical activity
5. Provides a framework for discussion and incorporating a wide range of views
6. Focuses attention on longer-term considerations
7. Indicates where a reorientation of technical resources is required
8. Provides a permanent record for subsequent reviews.

The form of the strategy statement will vary with the nature of the industry and its technology. Its main elements, however, are likely to include:

Financial The allocation of the technical budget between the various categories of expenditure.

Personnel Identification of recruitment and training needs to meet the requirements of the future. This might involve the setting-up of a group in a technology new to the company.

Products General specifications of products for introduction over the strategic planning period; target market segment – specialist or high volume; incremental improvements versus new designs.

Processes Process improvement versus new processes; processes versus products.

Organisation Changes in organisation structure, management etc. as appropriate for the strategic objectives.

Systems Changes in managerial systems, e.g. to exploit the potential of information technology.

Innovative stance The balance or choice between an offensive or defensive strategy.

2.14 The next stages

The strategy itself is only the starting point for engineering management. Its aim is to provide a general framework for *where* resources should be allocated. The next stage, the planning, is concerned with *what* should be done; this will usually involve the evaluation and selection of the projects whereby the strategy will be achieved. A successful outcome, however, depends upon *how* it will be done. Running through this chapter has been the theme that the main concern of a strategy is to identify major areas for change. There is a vital difference between recognising the need for change and its implementation. Thus the formulation of a technical strategy must be regarded as only a starting point for what might become a wide-ranging programme for managing technical change.

Chapter 3
Technical progress and engineering decisions

3.1 The need

The role of the engineer or technologist in industry is to apply science for the satisfaction of human needs. This may take the form of entirely new products or of a reduction in the price of existing products through improved design or higher manufacturing productivity. In this context it is important to draw a clear distinction between science, which is concerned solely with the acquisition of new knowledge for its own sake, and technological innovation, which is the process by which this knowledge is harnessed for a useful purpose. A study of the history of science shows that delays of several decades between these two processes are not uncommon.

There are a number of reasons why new scientific knowledge is not exploited immediately:

1. There may be no industrial incentive to accept the risks inherent in technological innovation. Companies in a mature but relatively profitable industry may be averse to making the investment and accepting the risk of changing traditional practices when they can see no threat to their established business.
2. The step from knowledge to application may have to wait until an individual with insight or creativity sees in it a potential which has not previously been noted or appreciated.
3. The release of the potential may depend upon advances in other technologies. For example the jet engine was not an economically viable proposition until developments in metallurgy enabled the manufacture of turbine blades which could operate at high temperatures.
4. The market potential may be difficult to assess.

It is the technologist who must bridge this gap. To do so he must be aware of scientific advances and must identify how they can be harnessed to meet a market need through new products or manufacturing processes. This is a challenging task since he must match his technical insights and creative ability to an understanding of what will sell in the market place. The processes by which this is brought about are often described by the terms *technology push* and *market pull*; their relative merits are a subject of much debate in the literature.

Technology push refers to the development of radical new products, emphasising their technological merits. Critics of this process cite the many costly projects which were stillborn. But it must be recognized that there are real problems of identifying whether there will be a market, and if so its size, for entirely new products. At the time the first television sets were designed it was by no means certain that there would be a market for the new medium. In this instance the innovation gave birth to a large new industry. This contrasts with the market failure of the video telephone following a heavy investment by AT & T of the USA; the product was exhibited at the 1964 World's Fair. It is doubtful whether any form of market assessment can provide a sufficient basis for making such decisions. Today similar question marks must hang over the size of the market for satellite or cable television systems. Thus the major successes of new technology and also the most costly failures have resulted from projects that can be categorised as technology push. Nevertheless such risks must be taken if the full potential of new technology is to be realised. This demands the faith and enthusiasm of an innovator within an organisation that is prepared to support him. In all caess, however, ultimate success must rest upon the development of a market for the product at a price for which it can be manufactured. Thus it is incumbent upon the developer to assess the market and financial implications, although it must be accepted that this cannot be done with any high degree of certainty.

Market pull, by contrast, results from the detailed research of what the market demands. This provides guidance about the attributes of the product perceived to be important to the consumer, which the technologist must attempt to satisfy. Since market research is unlikely to uncover entirely new market requirements it usually leads to less radical and incremental developments of existing products. Although this approach can be expected to result in a lower level of market failure, many of the potential rewards from new technological advances may be sacrificed.

It can be seen that both technology push and market pull have a place in the application of new technology. In the final analysis, however, it is the market place which will be the arbiter of success and will determine whether the investment in technology and the risks associated with it have been financially justified.

Whether he is involved with technology push or market pull the engineer is concerned with the application of advances in technology, sometimes based upon scientific principles which have not been used previously, but more frequently by developing extensions of existing technology. In doing this he must:

Capture new knowledge   This involves a wide-ranging monitoring of scientific and technological progress, often involving the combination of advances in complementary disciplines.

Forecast time scales Since the development of new products often involves long time scales for R & D and design, it is necessary to evaluate the rate at which new knowledge can be realistically incorporated in product or process designs.

Assess the market To do this he must identify the product attributes of most importance to the customer, and the rate at which the market may be expected to grow.

There are a number of approaches, discussed later in this chapter, which can assist the engineer in performing this task. It must be stressed that these techniques can never be a substitute for his technological and market judgement; nor can they eliminate the uncertainties inevitably associated with innovation. Thus the engineer should be aware of and use these techniques when they are appropriate, as long as he remembers that although they can assist his judgement they can never replace it. Futhermore, it must be stressed that an engineering project cannot be assessed in isolation from the business environment in which it is to be developed. Although a project may be technically feasible with a high probability of market success, it may be of little value if it does not fit the organisation's corporate strategy, demands more investment than the company has access to, or involves a greater risk than it is prepared to accept. In this context there are a number of questions which the engineer should address when he is evaluating the implications for the business of an improvement of technological performance:

1 Does it substitute for an existing technology in current products? If so, what is the likely time scale?
2 Will it make obsolete the current product line? If so, how long will it take?
3 Can it enable products to be developed which substitute for those not currently manufactured by his company? This might provide an opportunity by invasion and the basis for a technology-based corporate diversification.
4 Could it lead to the development of entirely new products to provide a unique competitive advantage?
5 How might it affect the total size of the market measured as either volume of output or financial turnover.
6 Might a relatively small incremental performance improvement have a significant impact upon the company's competitive positon and its market share?
7 Will the nature of the market and the potentail purchaser be changed? The pocket calculator made the slide rule obsolete but appealed to a much larger market and is sold through different distribution channels.

3.2 Historical background to technical progress

In order to understand where we stand today and the future directions of technological advance, it is useful to examine briefly the history of industry over the past two centuries. This reveals a number of distinct phases, or industrial revolutions, which have had a profound impact on business success and the focus for technical activities (Fig. 3.1). The technologies which have been the motive power for progress in one phase have often played a relatively small part in the subsequent phases. Where companies have been unable to adapt to these changes they have frequently gone out of business.

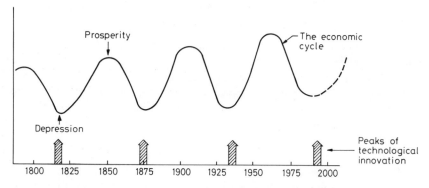

Fig. 3.1 *The long-wave economic cycle and technological innovation.*

Industry as we know it today commenced with the first of these industrial revolutions in the early part of the nineteenth century. This was founded upon the harnessing of new power sources based upon steam to replace those of animals and people. Apart from the notable exception of the railways the great majority of the resultant innovations did not lead to new consumer products. In today's terminology they were process innovations. The most significant development during this period was the growth of the textile industries in the north of England, producing large volumes of cheap cloth which was not very different from what had been manufactured by hand in earlier periods.

Towards the end of the nineteenth century the emphasis shifted to the expansion of the basic chemical, steel and heavy engineering industries. The electrical industry was also established towards the end of this period.

More recently we have seen the emergence of the mass production of consumer goods, with an explosive growth after World War II. This was a time for the rapid introduction of new products, a high proportion of which replaced activities the individual had previously performed for himself manually. In terms of radical new products this phase is drawing to a close, for it is difficult to conceive of many physical activities we now perform for ourselves which could be usefully mechanised and for which a product does not now exist. The demand for many of these products in the developed world is now largely saturated. As a consequence the consumer product industries are satisfying a replacement market with products which, for the most part, are only marginally better than the models they replace. These industries continue to exist, but on a smaller scale, with the prospect of little scope for further growth.

Technical progress and engineering decisions 57

Today we are witnessing yet another revolution based upon electronics and information technology. As in the past this current phase is based upon different technologies, and to a great extent is being exploited by companies which have only been in existence for a relatively short time.

Can these developments be fitted into a pattern? In the 1930s the Russian economist Kondratieff suggested that there was a 50 year cycle in economic growth – the long wave – followed by depressions (1830s, 1880s, 1930s). In the post-war growth period this theory was discounted, but it has been revived in recent years with the recessions of the 1980s following approximately half a century after the depression of the 1930s. Later economists (e.g. Mensch) have linked these long waves with technology. Mensch assessed the number of major technological innovations occurring at any time and noted that there was a rapid rise in their incidence, more or less coinciding with the depth of the depression. The rationale for this is as follows. The generation of new knowledge – science – progresses at an even pace but there is little incentive to apply it when the economy is growing. However, around the depth of the depression there is high unemployment, the traditional industries are in financial trouble, and entrepreneurs emerge to develop new activities exploiting the untapped potential of existing scientific advances. In many cases this occurs outside the established companies through the formation of new ventures.

In Chapter 2 the industry/technology life cycle was discussed. It will be observed that according to the long-wave theory these cycles are bunched in time, leading to the Kondratieff economic cycle.

Although the proof of this theory is inconclusive, the evidence of the 1980s seems to support it. We are currently experiencing the relative decline of established manufacturing industries, the growth of many new ventures based upon electronics and microelectronics, and the emergence of a host of new products in the field of information technology. Furthermore, this is largely being exploited by new venture companies or larger companies that have grown rapidly in the past 10 to 15 years (IBM, Texas Instruments). Microelectronic and information technology have an application in every industry and are the dominant technological influence on business at present. Nevertheless there are other science-based technologies which cannot be ignored, such as new materials (e.g. the engineering application of ceramics) and biotechnology, which may well be the driving force for the next long wave.

The implications of the above for business and the engineer are:

1 The changes we observe today seem to fit an historical pattern.
2 Companies in established industries must apply the new technologies if they are to survive at a time of transition, either by incorporating them in their existing products or by business diversification.
3 The contribution of technology in business is of critical importance during the economic growth period, implying the need for increased R and D budgets and a greater corporate role for the engineer if the large company is to survive the transition.

58   Technical progress and engineering decisions

4 The threat from small new venture companies cannot be dismissed.
5 The assessment of new technology and the ability to forecast technical change is essential in order to identify the opportunities and threats and to time the introduction of new products.
6 The individual engineer needs to gain an understanding of the new technologies, irrespective of his original discipline, if he is to discharge his corporate responsibilities.

3.3 The determinants of technical progress

Before discussing how engineers can improve their decision-making by forecasting the future progress of technology, it is necessary to examine the causes of technical advances. It must be remembered that what concerns the engineer is the rate at which a new technology (or products based on it) diffuses into the economy – an indication of the size of the market at times in the future. As we have seen already, this may bear little relationship to the rate at which new scientific knowledge is generated.

Examination of graphs of past technical progress would often seem to suggest that this occurs almost inevitably, following a smooth path when a technical parameter is plotted against time. This is frequently the case. However, before extrapolating the past into the future we must ask the question: what causes this progress? It is a consequence of the work of engineers and technologists who are paid and equipped by industrial companies or government agencies. These organisations have some tangible objectives without which they would not make these resources available. In industry this is the aim that the work funded will lead to a marketable product; in government there may be wider social objectives. If no foreseeable practical application can be identified these resources will not be allocated and there will be no progress in the technology. Thus any examination of future technological advances must be founded upon an understanding of the market from which the stimulus will originate.

The attributes the user requires from a product are often complex and may not always be those which the engineer's logic might indicate. Let us take as an example the factors which might have to be considered when assessing the future demand for solar energy in Europe. At first sight this may appear to be a relatively simple calculation provided the engineer can quantify the relevant costs and benefits; from these he can evaluate the net present value of an investment in solar energy to the consumer. But we must ask whether in making his decision the consumer will evaluate it using the same criteria. Who is the consumer? Is it the builder who incorporates solar heating in new houses, thereby adding to their price? Or is it the house owner who converts his heating system? In the latter case, is he concerned with cost, comfort, reliability or independence of external sources of energy supply? How does he relate a capital expenditure to an annual expense saving? At what interest rate would he pay to borrow the capital? It can be seen that the problem is much more complex than originally thought. Furthermore, it typically leads to a much less optimistic prospect than that indicated by the engineer's initial evaluation. There are numerous instances where developments have been undertaken on the basis of a technical analysis without a clear understanding of what motivates the market, leading to an over-optimistic assessment of its potential.

Thus in assessing a technological potential either in its own right or as a business opportunity it is necessary to gain a view of what *will* occur, not what *can* occur or what *should* occur. Of course it will not happen if it cannot occur, but it will not necessarily happen because the technological feasibility exists. The factors involved may be complex, involving economic, social and political as well as technical considerations. Furthermore, some of these factors may be difficult if not impossible to quantify. The engineer must take care not to ignore or underrate the importance of these qualitative considerations. They are real even if they cannot be incorporated into a mathematical model.

3.4 The elements of technology forecasting

The first and perhaps most important step in assessing the opportunities which technology might present to the organisation is to *identify* those technologies or technological developments which might be of significance. In other words the engineer must determine *what* be should be considering. If he is not taking into account the appropriate technical phenomena or potential developments, he cannot identify the areas where the organisation should focus its attention. This is not as easy as might appear at first sight. Inevitably most engineers will naturally suffer to a degree from 'tunnel vision' in that their attention is likely to be dominated by the disciplines in which they have been trained and the products the company had traditionally manufactured. But new technology inevitably involves some elements of change, either gradual or radical. The new technology may completely replace existing expertise, as with the case of microelectronics in the electromechanical industry, or may be incorporated into existing products, as with plastics and ceramics in the automotive industry. As we have stressed earlier this cannot be done unless the engineer has an understanding and appreciation over a wide range of technological and scientific advances. Normally this does not require a deep knowledge for, once the potential has been identified, experts or consultants in the appropriate areas can be employed to evaluate the implications in greater detail.

The first step of identification, although based upon knowledge, does not follow automatically; it requires insight and creativity. The new technology may not, however, lend itself to immediate application. Most products incorporate a number of technologies. The exploitation of the new technology may not be possible until there has been an advance in one of these associated technologies. It is, therefore, necessary to *monitor* these other technologies so that when they have reached the necessary performance the combination can make possible the product development. The dependence of the early jet engines on metallurgy for high-temperature turbine blades has already been noted. In this case one could postulate a situation in which an engineer had identified the theoretical advantages of jet propulsion, appreciated the limitations imposed by current materials and established a monitoring system so that he would receive timely information on when these limitations had been overcome. As a consequence he would be aware immediately the jet engine was a practical proposition. This would put him in a position of initiating development in advance of competitors. If a company has a technologically offensive strategy it is essential to be the first in the field. A few months lead can be critical for commercial success. A monitoring system can play a crucial role in establishing this lead.

So far we have considered what the engineer should be assessing; this is essentially a qualitative exercise. Attention must now be directed to the quantitative measurement of the technology. In some cases this is straightforward; for example, with jet engine turbine blades the measure of performance presents little problem in that it would be a characteristic such as the creep resistance of the material at various temperatures. However, one must not lose sight of the attributes required by the customer. An airline, for example, is not concerned with the specific fuel consumption of an aeroengine. The attribute it values is the cost per passenger mile of the total system – that is, a measure of the aircraft's productivity. Thus an engine with a high specific fuel consumption (SFC), the jet engine, yields a lower cost per passenger mile than a piston engine with a lower SFC because the higher speed of the jet leads to improved aircraft productivity. A similar problem arose with shipbuilding, where the industry measured its performance as cost per tonne of ship, failing to realise that the ship owners were prepared to pay more for ships if this led to lower operating costs. Another problem occurs where incremental improvements in some feature of technological performance are not apparent to the purchaser. The manufacturers of lavatory disinfectants measured their performance by the ability to kill microbes and in some cases invested in increasing this ability. However, the user is likely to be confident that the product will acheive an adequate performance as a disinfectant; the attributes which sell the product today are colour and perfuming, not necessarily those uppermost in the minds of the technologists. Once again one notes the dominance of the market place and the need to identify the performance attributes important to the customer and to measure them wherever possible. An obsession with purely technical performance criteria may often lead to costly development which is not relevant to the customer and consequently of no value to the business.

The third element to be considered is *time*. As a generalisation it can be stated that many engineers do not pay sufficient attention to assessing the rate at which a technology is evolving. All engineering decisions come to fruition at some time in the future; in some industries this may be upwards of 10 years. Thus in drawing up a specification for a product or process it is necessary to forecast what will be technically feasible at that time and also make a judgement of the probable competition. In doing this it is often necessary to consider the total system within which the product will operate. An example of the importance of the total system is provided by the VC10 aircraft, which was considered to have a competitive advantage over the Boeing 707 in that it could operate from high-altitude airports in Africa. However, by the time the VC10 was in service these runways had been lengthened to enable the operation of the Boeing 707, thereby making irrelevant what had at first been thought to be the VC10's competitive advantage. This error in anticipating future developments led to the design of an aircraft which proved to be a commercial failure, not because it failed to meet its technical specification, but because the specification was inappropriate to the needs of the market at the time the aircraft came into service. Frequently new products are initiated by comparing them with the performance of the competitor's current offerings rather than with what will be available in the future.

If the progress of technology was a succession of random jumps or discontinuities it would be difficult if not impossible to do other than guess in making these decisions. Fortunately this is not the case. As we shall see later there are patterns of past progress which can be used to make technological forecasts. Thus it can be seen that:

1 Consideration must be given to the total operating system. It is future market needs which should be the determinant of the technical specification. A perfect technical solution to the wrong problem spells commercial failure.
2 Failure is most frequently the consequence of overlooking a *big* signal. Whilst it is always important to have good detail design, it is of little value if the specification is not aligned with long-term market needs.
3 Timing and an assessemnt of the perforamnce level of technology into the future is essential. Valuable assistance can be gained by using the techniques of technology forecasting.

Having established future technological potentials and the attributes required by the market (i.e. what we should be looking at), the parameters by which these factors can be measured (i.e. a quantification of technical progress), and the development of these parameters through time, we have a basis for making our decisions. But we know that there are likely to be a number of developments which will introduce errors into our forecasts. It is a truism that all forecasts are wrong. These uncertainties must be built into our assessments by attaching a *probability*. Some factors will be associated with a greater degree of uncertainty than others in the judgement of the engineer making the assessment. This must be reflected in the forecast by associating it with a probability distribution; inevitably this must be subjective.

Thus a forecast statement might take the form: by 1995 (time element), 70% of private houses in Europe (quantification of level) will have solar heating (statement of the event or phenomenon) with an 80% probability (probability statement).

Critics of forecasting question its validity because of the incidence of discontinuities. It must be accepted that there will be, from time to time, events whose time of occurrence it is not possible to forecast. Nevertheless in many cases such events can be foreseen. A good example of this type of event is provided by nuclear power; the possibility of a major disaster could be foreseen, although there was no way in which the date of the occurrence – Three Mile Island or Chernobyl – could have been forecasted. However, once the event has occurred it may have so significant an impact upon the industries or technologies involved that consideration of the consequences cannot be ignored. In such instances all that the business planner can hope to do is make contingency plans so that prompt action can be taken. Additionally, of course, there will sometimes occur events which no one is likely to have foreseen; in such cases it is impossible to take any action in advance. Although major discontinuities do occur their frequency in most industries is sufficiently low not to invalidate a forecasting activity for most of the time. Furthermore, the possibility of these events, once identified, can be taken into account in both business and technological planning.

Four elements which must be incorporated in any forecast have been identified, namely:

The event or phenomenon, i.e. a qualitative identification of what the forecaster should be addressing – a new technology, a market attribute or an event

A measure of performance, being the parameter by which the attribute can be assessed quantitatively

The time scale, linking a performance level of the parameter and a date in the future

A Probability assessment, albeit subjective, to reflect the inherent uncertainties of future developments.

It must be recognised that in any decision, be it business or technical, a view of the future must be taken. The purpose of a forecast is to ensure that this view reflects the best judgements that can be taken with the information available at the time it is made. Uncertainty can be reduced even though it cannot be eliminated.

3.5 The inputs to forecasting

In the previous section the elements of a forecast were identified. If one of these elements is missing the forecast does not provide an adequate basis for a managerial decision, or may be ambiguous. It is now necessary to examine the inputs to the forecasting process.

In our thinking there are always a number of *assumptions* which form the basis of our understanding of how any system works. These assumptions are high-probability factors which are normally taken as 'givens'. They result from past experience and provide a model for decision-making. These assumptions must be questioned since they may not sufficiently reflect the condition of the future. For many years there was a relationship between GNP and energy consumption. On the assumption that this trend would be maintained, forecasts were made for long-term energy requirements which have proved to be greatly overestimated owing to the impact of the 1973 oil crisis. This, manifesting itself in higher energy prices, stimulated conservation and a departure from the previously well established relationship. The result was a cessation of orders for new power stations, with severe consequences for companies such as NEI and Babcock. A similar relationship existed between GNP per head and the consumption of meat. In this case the established relationship was changed by increasing public concern with health, which led to a drop in the per caput consumption of meat. In the first example the stimulus for change was economic, in the second social. Thus we note that:

1 Assumptions should be identified and assessed in relation to the future when they may no longer be true.
2 They should be noted and made explicit.
3 The factors which invalidate previous assumptions frequently arise from economic, political or social changes. Nevertheless they can have a profound impact on businesses and the technologies associated with them.
4 Mathematical forecasting models always incorporate assumptions. In a number of instances these models have been used without modification to their programs to reflect the changed relationships, thereby leading to unnecessary forecasting errors.

The next input to the forecasting process which must be considered is the *quality of the thinking* applied to the evaluation of the future. In discussing the outputs it was stressed that the starting point must be the identification of the appropriate phenomena or events which will shape the future environment of the business. This demands the ability to take a wide view across the social, political, economic and technological environments and to spot significant relationships between them. If this is not done the application of the forecasting methodologies will lead to a result which is 'precisely wrong'. The people who have the experience and insight to take such a view will normally hold senior positions. Furthermore, it is unlikely that many engineers will have the knowledge or time to monitor those developments outside the technological field. Thus bridges must be built to those who do have expertise in corporate planning, marketing and finance, so that there can be a true meeting of minds to explore the interrelationships. In this context the contribution of top management with their wide range of contacts external to the company should not be underestimated.

In order to quantify it is necessary to have *data*. These must of necessity relate to the present and the past. Unfortunately the quality of data is usually poor. At one extreme we have demographic statistics which for most practical purposes are sufficiently accurate. At the other extreme is the social area, where usually it is only possible to measure indicators for the attitudes it is desired to measure. For example, one may wish to study the co-operativeness of the workforce; although this is a valid concept it cannot be directly measured, and indicators such as the incidence of strikes, absenteeism or sickness have to be used. Technological data fall between these two extremes. A major problem is that technological historical data are not normally collected or recorded on a regular basis. Thus when ones wishes to track the progress of a technological parameter (for example the maximum tensile strength of steel from 1935 to the present), there may be no readily available source. Often considerable research is required to locate the figures desired, and even then they may contain ambiguities in either the definition of the parameter or the conditions and method for its measurement. This may appear to restrict the value of the data for forecasting. However, for most engineering decisions one is looking for big messages; these should not be highly sensitive to the quality of the data, provided it is sufficiently accurate for the establishment of a trend. One must always keep in mind the decision that has to be made. For example, in relation to building a plant for a new product the alternatives may be to cater for a demand of 75 000 or 100 000 per year; the choice will not be between 98 000 and 100 000. Thus:

1 Data are needed to provide a quantitative measure for the phenomenon in which we are interested; they are needed so that graphs can be plotted to establish trends.
2 In many technologies and industries historical data have not been recorded, so the forecaster may have to acquire them from his own researches.
3 Forecasting is concerned with long-term decisions which usually involve major investments; shortcomings in the quality of the data are likely to be insignificant in relation to other uncertainties. Sensitivity analysis can be helpful in this context. If both a high and a low interpretation of a trend line indicate the same decisions a greater degree of accuracy is unnecessary; this is often the case. If, however, the two interpretations lead to different conclusions it must be accepted that the forecasts are of little practical value except to underline the high level of uncertainty and risk associated with the decision.

Fig. 3.2 illustrates diagrammatically the points discussed so far. The stimulus for the forecast may come from:

(a) The desire to explore the business impact of an evolving trend
(b) The insight of a manager wishing to substantiate or reject an idea
(c) A trend or relationship emerging from a forecasting exercise
(d) The need to evaluate the appropriateness of a business or technological strategy or decision.

Before proceeding to describe some of the forecasting techniques, it is worth repeating three points of great importance:

Technical progress and engineering decisions   65

1 The output from a forecasting exercise can only reflect the quality of the inputs. The sophistication of a technique should not be allowed to engender a sense of false accuracy.
2 Few if any decisions can be taken solely as a result of considering technology in isolation from economic, social or political influences. This implies that forecasting for technology cannot be conducted without inputs from members of other functions in the business.
3 Forecasting in isolation from the decision-making process is of little value.

Trends or events →	Inputs	→ Methodology →	Outputs	→ Purpose
Economic	Assumptions		Phenomena, attributes, events	Business or
Technological	Insights	Forecasting Techniques	Measures, parameters	technical decisions
Social			Time	
Political	Data		Probability	

Fig. 3.2 *The forecasting process.*

Large engineering companies may have a small group of technology forecasters. Most smaller organisations are either ignorant of the contribution forecasting can make or do not consider the employment of forecasters to be financially justifiable. The author's view is that a knowledge of the techniques of forecasting should be possessed by all engineers. This knowledge can be easily acquired since in most cases the simplest approaches are usually the most valuable. A specialist group can become remote from the mainstream of decision-making and can engage in generating forecasting which may be of interest but of little practical value. In the largest companies where a specialist group (usually multidisciplinary) may be justifiable, it is essential that they are integrated into the decision-making process. However, there are two areas where the engineering manager may require help: the acquisition of data, and assistance with some of the techniques. Thus it is essential that an organisation collects historical data in relation to the environments of the business which are available to all managers. The technical assistance in forecasting techniques can be obtained from consultants or by the training of a few people who can be called upon to give specialist advice when required. Forecasts made by the decision-maker himself are likely to have a greater impact and credibility than those provided by a specialist group.

3.6 The causal model

Technical advances do not occur in isolation. They are made to happen by the efforts of engineers and technologists. If the resources to fund them are not made available then there will be no progress. The allocation of finance is made because it is judged that it will result in a product or process for which there will be a market. But the market itself is determined by a variety of factors. For example, the customers' needs might be satisfied by products based upon different technologies or systems; the motor car is in competition with public transport, and the petrol engined car is in competition with the diesel and perhaps the electrical car at some future date. The size of the market is influenced by purchasers' disposable income, itself influenced by government economic policies, which are determined by the political party in power. Social trends may also have an influence sometimes difficult to assess; environmental concerns or trends to health food can have an important impact on some manufacturing industries.

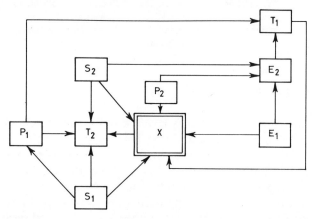

Fig. 3.3 *Influence diagram showing some of the factors relating to development of domestic solar heating.*

Some of the variables

X	= Attribute being forecast	e.g. Domestic solar heating
P_1	= Political event	e.g. Anti-pollution laws
P_2	= Political event	e.g. Opec oil price rise
E_1	= Economic factor	e.g. Personal disposable income
E_2	= Economic factor	e.g. Household energy cost
T_1	= Technological factor	e.g. Cheap solar cell
T_2	= Technological factor	e.g. Cheap wave/wind power
S_1	= Social factor	e.g. Environmental concern
S_2	= Social factor	e.g. Desire for personal energy independence

Thus we have a complex set of interrelationships comprising technological, economic, social, political and environmental factors. If all these factors can be identified and their interrelationships determined it is possible to construct an influence diagram of the type shown in Fig. 3.3. It will be noted that the direct influence on the technological development of interest can be positive (i.e. make it more likely) or negative (i.e. make it less likely). The influence diagram shows in a qualitative way the causality determining the progress of the phenomenon of interest. In theory it should now be possible to develop a computer model.

Although models of this type have a great appeal in that they are based upon definable causal relationships, they are not widely used as a practical forecasting tool for technology within industry. There are a number of reasons for this. A major problem is the complexity of the model if all the interrelated factors are incorproated into the influence diagram. In order to simplify it into a usable tool the forecaster must exercise his judgement. Often it is difficult to establish the quantitative relationships between the variables; some of them do not lend themselves to quantification, and a serious weakness of many models is that social and political influences are omitted because they cannot be quantified.

Most of the useful techniques for forecasting for technology are observational. Over a period the technology will progress in a way which can be measured

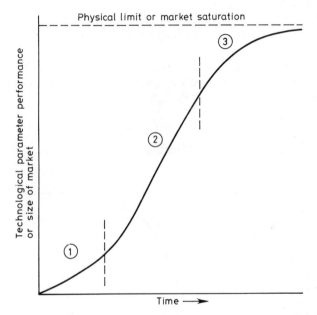

Fig. 3.4 *The S curve of technological progress:*

even if the contributions from the various influences cannot be isolated. This past progress can be extrapolated into the future and will provide a reasonable basis for a forecast provided there is no significant change in the importance of any of the influencing factors. Thus although it is normally impracticable to use a causal model for the reasons outlined above, it is still possible to make forecasts by:

Observing past tends and extrapolating into the future
Preparing an influence diagram which yields an understanding of the underlying causes of the phenomena in which we are interested
Monitoring the major influences to identify any significant changes which might invalidate the trend

Modifying the trend forecast in the light of the changes noted; in most cases this must of necessity be largely subjective.

3.7 Patterns of progress

Forecasting for technology would not be possible if it advanced in a random fashion. Fortunately, an examination of a large number of technologies shows that there is a characteristic growth curve of S shape (see Fig. 3.4). In most of the literature the parameter is drawn dependent on time. However, the causal relationship is between technological progress and, not time, but the cumulative investment that has produced that progress. In recent years the consulting company McKinsey has researched a number of industries where it is possible to establish the cumulative investment to derive S curves. However, in many cases the only data available relate to time. Fortunately in most examples these also produce S curves which can be used for forecasting, provided it is not forgotten that there is no causal relationship between progress and time; this is one reason why the monitoring of the influence diagram noted in the previous section can be important.

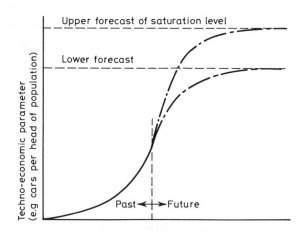

Fig. 3.5 *S curves for high and low forecasts of saturation level.*

The upper level of the curve is determined by some limit to which it is asymptotic. Where a purely technical parameter is considered this upper limit is set by the physical laws of nature and is clearly defined. The S curve does, however, apply to a considerable number of techno-economic or market determined growths, for example car ownership per head of population. In these circumstances the limit is the saturation level for the product, which cannot be established accurately in advance. In such cases it is customary to set two levels, high and low, for the saturation level and to derive two S curves (Fig. 3.5). The differences between the curves for the alternative saturation levels may appear significant, but are often of less importance when considered in relation to the business decisions based upon them. For the important indication is the early warning the graph gives of the approach of saturation at a time when the total market is still growing strongly. In this context it is the realisation that the period of rapid growth will come to an end within 8 to 10 years that is important rather than an accurate forecast of whether it is 8 or 10 years. It has been noted before that the greatest value to be derived from forecasting is the early warning of *big* changes. Many industries have over-invested at this critical stage in their evolution because attention has been focused on the past, when growth has been rapid, rather than on the future, which marks the approach of saturation.

There are a variety of mathematical relationships giving a curve of S shape e.g. the logistic, the Gompertz. Although their relative merits are debated in the forecasting literature their variations are unlikely to lead to forecasts sufficiently different to affect the business decision. In practice the logistic curve, a symmetrical S about the midpoint, is normally used.

Readers will have noted the similarity between the industry lifecycle (see Chapter 2), the product life cycle and the S curve of technical progress. The industry life cycle is closely dependent upon the S curve but is not normally quantified; it is used as a qualitative concept to aid the understanding of the evolution of an industry. The product life cycle is usually considered in relation to a shorter-term product model (e.g. VW Golf) whose life is largely determined in relation to newer models, which may not differ from it widely or employ new technology. The S curve of technology is concerned with the longer term; is quantified; relates to a specific technology or generic product type; and is used to forecast future trends as a basis for major decisions in relation to the business or its technology.

Figs. 3.4 and 3.5 show S curves which are partially formed as a result of past growth. Often, however, the technologist is faced with assessing the future potential of a new technology before it has established itself. One can identify three questions to which he would like to have answers, which of necessity must be based upon forecasts.

How long will it be before the technology becomes significant?
This is the early part of the curve, which is the most difficult to forecast. Sometimes it may be decades before a feasible development establishes itself. As long ago as the late 1950s articles appeared with titles such as 'Life in the automatic factory', but it was only in the late 1980s that the development of robotics brought closer the initiation of largely unmanned manufacturing units. The delay was mainly due to the absence of a technologically economic solution; it was a feasible concept awaiting the appropriate technical advances. Another example is provided by carbon fibre composite materials. As with most products the unit price is highly dependent upon the output volume. At low volumes the high price is only justifiable for specialist applications. A theoretical low price

at very high production volumes cannot be realised in practice unless a rapid growth in the market can be assumed. This did not occur in spite of high initial expectations. Only in the late 1980s is the demand increasing with its use as an aircraft structural material. Frequently the early applications of a new technology are for a specialist use where the customer has no alternative; later it expands into larger markets where it substitutes for an existing product as the

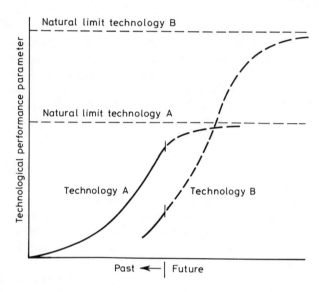

Fig. 3.6 *Technological substitution.*

cumulative production forces down the unit price (see the experience curve, Fig. 4.6).

In other cases the lead time may be short if the merits of the new technology meet a mass market demand; this is often the case with new pharmaceutical products. In evaluating a technology in these early stages of development the forecasting must rely heavily upon subjective techniques such as Delphi, based upon the judgement of experts (see Section 3.11), in order to gain a view of the pace of progress. Nevertheless, since the ultimate determinant of this progress lies in the market place, a techno-economic analysis associated with market research can be of assistance. Although the term 'market research' has been used it is important to draw a distinction between what is relevant here and the detailed survey of the market normally associated with the term. In this early stage it is the technologist who must assess the attributes of the technology and match them in so far as he can with potential market needs. Little is gained by surveying potential customers, for they will usually give a favourable response when not faced with a buying decision.

The aim of the forecast is, of course, to lead to a business decision, which is neither too optimistic nor too pessimistic, of the growth prospects. An overoptimistic view, not infrequent in small high-technology companies, results in the launch of products in which the technology is not adequately proven or the cost is too high. The opposite view could surrender the competitive advantage to another company. The key to success is matching the technology and the market at the right time; forecasting, although not able to make the match precisely, can help to steer a path between wishful thinking and scepticism.

At what rate will the new technology advance?
This question is directed towards evaluating the slope of the curve during the growth phase. The answer will give guidance on the timing and scale of the investment that may be necessary. It is dependent upon a number of factors:

Product performance If the new technology offers a benefit not possible from earlier technology it may be expected to diffuse quickly into the market. Alternatively it may offer considerable cost advantage over its predecessor even though its technical performance is little different.

Product price A low-priced product does not present the purchaser with a major buying decision. If the product costs only a few pounds he has little at risk and will be more inclined to move from the old to the new. Thus the electronic watch captured the market within a few years, whereas the diffusion of float glass took decades. In general process innovations involve major investment decisions and their rate of introduction is lower than for new products unless the latter are expensive.

Investment in the earlier technology There are two major reasons, both financial, which reduce the rate of diffusion. The first relates to sunk costs (see Chapter 6); although purchase of the new technology may be economically sound, there may be a reluctance to accept this rationale when a larger investment has been made in its predecessor. In some cases this will slow the rate of diffusion; in others the potential will be exploited by companies new to the industry which are not saddled with the consequences of the past investment decisions. The second factor relates to cash flow and the financial resources of the company. However desirable the new technology may be thought to be, the lack of funds with which to purchase it can override all other considerations. The larger the investment, the more likely is this to be a determinant of adoption.

Availability of infrastructure Many products are dependent upon a total system beyond the control of the product manufacturer. If this infrastructure is not available the market may not exist or may be severely restricted. Satellite TV receiving aerials will not be bought in the absence of a broadcasting service. The lack of diesel pumps on garage forecourts was initially a deterrent to the purchase of diesel cars.

When will saturation be reached?
This aspect has already been discussed briefly. It can be a critical business decision since in many cases the manufacturer has a major investment in the

72   Technical progress and engineering decisions

existing technology. Where a techno-economic factor is involved, say car ownership, it may be necessary to consider a business diversification if growth is to be maintained. In the case of the approach to a technological limit it normally implies the satisfaction of the existing market, which may still be growing, by products incorporating a different technology; we shall return to this aspect later.

3.8 Technological substitution

As a technology approaches its natural limit the scope for further growth diminishes. This brings with it a rapid increase in the R & D investment required to achieve incremental improvements. In other words the productivity of R & D falls sharply if it is regarded as the increase in performance per unit of investment. This may still bring considerable financial benefits if it yields a competitive edge over the opposition. A good example of this is provided by the development of the float process for the manufacture of flat glass, where the new technology enabled the developer (Pilkington) to achieve domination of the industry worldwide although the benefits in terms of manufacturing cost savings were not large. However, in a mature industry where profit margins are low the high R & D cost for relatively small improvements in performance can be justified. Such developments are not common and are usually confined to manufacturing processes.

More typical is the emergence of a new technology with a higher physical limit (Fig. 3.6). It can be seen that within a few years the performance of technology B will exceed that of technology A, and that when it exceeds the physical limit for A it has an unassailable performance advantage. If cumulative investment rather than time was on the horizontal axis, it would be seen that the productivity of the R & D investment for B is consideraby greater than for A. This would appear to present an undisputable case for a company to plan to abandon products based upon A and to shift its technological focus to B. Frequently this does not occur and many highly respected major companies have suffered severe losses as a consequence. There are many reasons for this:

1   Managerial conservatism, a reluctance to change, and complacency based upon past success create a climate in which 'good reasons' will be advanced for disregarding or discounting the significance of the new technology.
2   A widespread ignorance of the dynamics of technological progress means that most general managers and many engineers are unaware of the processes indicated in Fig. 3.6. This indicates a need for engineers to familiarise themselves with this knowledge and to communicate it to others in their companies who do not understand these growth patterns or appreciate their significance.
3   The initial products incorporating the new technology may exhibit only a small performance advantage over the traditional technology. In some cases

the overall performance may be inferior and the technology only achieves success in a specialist market where it does have a competitive advantage. Without an understanding of the potential for future progress it may be dismissed as not presenting any threat to the company's established products.
4 Although the threat is recognised the response is to protect the existing investment by accelerating R & D expenditure on technology A or by improving the productivity of existing plant. In retrospect that will be seen as 'throwing good money after bad'. This investment may delay the crossover point where the old technology becomes hopelessly uncompetitive. This phenomenon is sometimes called the 'sailing ship effect', referring to the marked improvement in the performance of sailing ships after the emergence of steam. Major investments were made in the construction of new canals after the first railways were in service; some of them always operated at a loss until closure. RCA invested heavily to preserve its radio valve manufacturing plants after the development of the transistor.
5 The threat is recognised but there is not thought to be any urgency to adopt the new technology. This is a failure to appreciate the rate at which the performance of a new technology advances once it has entered the high-growth part of the S curve. For example, the proportion of electronic cash registers in the USA grew from 10% to 90% in the period 1972–6. National Cash Register, the market leader in electromechanical equipment, barely avoided bankruptcy through its failure to move to the electronic registers sufficiently early; it suffered a $140 million write-off of electromechanical plant only a year after the investment, and made 20 000 workers redundant.

This is one of the most important areas where technology can have a major impact upon business success. All those industries where engineers are employed are liable to experience this form of substitution, which may affect the product either in its entirety or in part. Indeed it is an important role of the engineer to effect these changes, although a precondition is the allocation of funds following a recognition of the need at the highest level of the company. Yet one observes that failure to act is widespread and that many market leaders, including those which have invested heavily in technology, have suffered considerable losses as a consequence. Within these companies there were often some engineers who had a clear vision of what should be done; in some cases they were totally ignored or only half-heartedly supported not only by general management but also within the technical function itself. In retrospect it is easy to see that big signals were being ignored – signals which were not sensitive to forecasting errors.

The individual engineer cannot be held responsible for these failures but he can play important part in attempting to reduce them. He alone can assess the significance of new technology. If he has failed to identify it and communicate it he must share the responsibility for the consequences for the business as a whole. In summary he must:

74   Technical progress and engineering decisions

1 Identify the critical technical or techno-economic parameters for business success
2 Establish the physical limit for the existing technology
3 Establish the trend for the parameter in the past and evaluate the potential, i.e. the difference between the physical limit and current performance
4 Identify emergent technologies, their physical limits and their progress to date
5 Assess the implications of 1 to 4 for existing and potential products employing the new technology for the company
6 Present his findings in a fashion which communicates effectively with managers who do not possess a technical background.

3.9 Attribute substitution

Many products possess several attributes related to market need. For example, those for the motor car include fuel economy, cost, reliability, durability, speed, acceleration, road holding and comfort. Each of these attributes can be taken separately and analysed in relation to possible technology substitution as discussed in the previous section. A possible example here is the ceramic engine which, because it has a higher physical limit for its maximum temperature of operation than metal, offers a higher thermodynamic efficiency.

It is not sufficient to examine each attribute separately; they must also be assessed in relation to each other. At any time the market will be segmented in relation to the weightings accorded to each of these attributes within a group of consumers. Thus there will be a trade-off between speed and economy; some customers will be prepared to sacrifice economy in the interests of speed, and others will take the opposite view. These trade-offs are necessary because of the 'state of the art' in the technologies involved, which advances with time; thus in the 1980s there are combinations of speed and fuel economy which could not have been achieved in the 1940s.

These considerations affect the engineer in a number of ways:

1 There will be a minimum level acceptable to all users. It can usually be assumed that this has been achieved before the first product has been marketed.
2 There will be a maximum level for most practical purposes. One might consider corrosion resistance for the motor car, where the upper limit could be an unlimited life. However, there are likely to be few purchasers who would require, and pay for, a life in excess of 15 years. This will set the maximum performance for which the engineer should design even when it is far removed from the physical limit; enough is enough.
3 Each of the attributes, represented by the appropriate parameters, will have a position on an S curve. Those near the physical limit or the useful maximum have little to offer from further investment in R & D, whereas those at an earlier stage have more to offer. This would indicate a shift of emphasis from

the attribute which has a low R & D productivity to one with a higher. In many cases the first company to see this opportunity and shift the emphasis of its R & D gains a competitive advantage.
4 From time to time some event will occur which affects the desirability of an attribute across all market segments. Thus the rise in oil prices in 1973 changed the trade-off weightings between motor car fuel economy and the other attributes.

3.10 Product substitution

The analysis of technology substitution provides information on when the performance of a new technology is likely to surpass that of its predecessor. This does not imply that there will be a sudden change in the buying habits of customers. The decisions facing the businessman are firstly of timing (when he should market the new product) and secondly of scale (how large a manufacturing unit he should establish). In discussing the S curve we have examined some of the considerations affecting the rate of diffusion of a new technology. However, the decision-maker desires evidence of what is actually happening before he will usually make an investment.

The initial developments at the beginning of the S curve are to a large extent a gamble, for it is extremely difficult to make any useful forecast for this period. The first applications will result from the specialist market, the enthusiasm of entrepreneurs, or speculative R & D on the part of the large company. For example, the diesel engined motor car was available for many years before it sold in any great numbers. The market for it grew slowly until the mid 1970s, when it embarked upon a period of rapid growth; it was at this stage that the major motor manufacturers were faced with the decision of time and scale of entry.

Fortunately there is a forecasting technique – the Fisher-Pry substitution model – which is a powerful tool for making these decisions. Examination of a large number of product substitutions indicated that the diffusion of the new product followed an S curve, and the decay of the old followed a reversed S (see Fig. 3.7a). It further showed that when the substitution reached about 5% it was likely to be maintained, and that the dynamics of the substitution were sufficiently well established to allow the remainder of the substitution to be forecast. The algebraic expressions of these relationships are as follows:

$$f = \tfrac{1}{2}\{1 + \tanh[\alpha(t - t_0)]\}$$

or

$$\frac{f}{1-f} = \exp[2\alpha(t - t_0)]$$

where f is the fraction substituted, which is half the annual fractional growth in the early years, and t_0 is the time when $f = 1/2$. When $f/1 - f$ is plotted on semi-logarithmic graph paper, this relationship represents a straight line from which f (i.e. the proportion of the new product) can be derived (Fig. 3.7b). The steps involved in carrying out the forecast are:

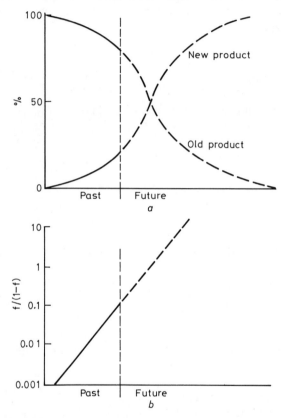

Fig. 3.7 *Product substitution: (a) substitution curves, (b) extrapolation using Fisher-Pry function.*

1 Collect data for existing and new product.
2 Calculate $f/1 - f$ and plot on three-cycle log-linear paper with 1 as the maximum. This is to enable a good straight line to be fitted to the points.
3 Replot the straight line from step 2 on three-cycle log-linear paper with 10 as the maximum and extrapolate.
4 Calculate f as follows. Let $f/1 - f$ at time t be y (i.e. $f_t/1 - f_t = y$). Then

$$f_t = y(1 - f)$$
$$f_t(1 + y) = y$$
$$f_t = y/1 + y$$

Technical progress and engineering decisions 77

For example, when $y = 1, f = 0.5$; when $y = 10, f = 0.91$. Thus by calculating f at dates in the future it is possible to draw the S curve for f.

5 Forecast the total market for both old and new (T).
6 Calculate demand for the new product at dates in the future, i.e. $f_t T$.

Examples of where the use of this technique derived accurate forecasts of future substitution growth of a product based on a new technology include: the open hearth furnace for Bessemer steel; detergents for soap; radial for cross ply tyres in the UK, 1969 (34%) to 1981 (93%); and the diesel for the petrol car in Europe, 1973 to 1986.

This is an extremely useful technique which is strongly recommended for use by all engineers in assessing the long-term market trends for a new technology-based product once it has established a toehold in the market. There are, however, a number of caveats which should be borne in mind. The substitution is a model of human behaviour in that it aims to forecast the buying decisions of individual customers. If one of the environmental influences changes, so might the slope of the straight line. The oil price rise in 1973 produced a dogleg in a number of these curves, e.g. the substitution of cellulosic manmade fibres by oil-based synthetic fibres, and rather surprisingly the substitution of monochrome by colour TV in the UK. Whereas the diesel for petrol car substitution held at a constant rate in Europe this was not true in the USA, where it suffered a severe setback following the failure of GM diesel cars in service – a circumstance nobody was likely to foresee.

Frequently a new technology segments the market. Thus there remains a market for soap and natural fibres, although a major part of their former markets has been substituted by detergents and manmade fibres. In such cases it is necessary to make a forecast or judgement of the proportion of the total market which will be substituted eventually and use this as the 100% level for the analysis.

How should these forecasts be used? It is known that in some cases there will be environmental changes that were difficult to foresee, thereby to some extent invalidating the forecast. It is not possible to say: 'This is what will happen'. On the other hand it is valid to state: 'In the absence of any new factors this will be the state of substitution based on past trends which have followed an established pattern'. The forecast can thus be used as an agenda for a discussion on future developments. The onus is placed upon those who disagree with it to put forward their reasons, which can then be rejected or substantiated. Thus it provides the basis for an informed debate from which new factors may be adduced in the light of which the forecast may be amended. At the end of this process the best possible view of the future will emerge, based upon past trends and the judgements, knowledge and insights of those within the company.

It must be noted that these forecasts relate to the total market and not the market share of the company, which depends upon internal decisions such as pricing. But armed with the forecasts the top management must decide whether to enter the market with the new product, when to do so, and on what scale.

78 Technical progress and engineering decisions

3.11 Forecasting techniques

There are a large number of techniques which have been developed for the forecasting of technology. They can be broadly categorised into exploratory and normative types. All the approaches describes in this chapter fall into the *exploratory* classification, in that they have been based upon an extrapolation of past trends into the future using patterns which have been shown to hold for a large number of past technological developments. These are the most frequently used approaches.

In *normative* forecasting the aim is to postulate or speculate about some possible future event or development and then attempt to trace a path from the present by which this state could be achieved. It is beyond the scope of this book to describe these techniques in detail, but three of those most frequently used will be mentioned briefly:

Delphi is a method to structure the views of experts anonymously using a questionnaire. For example, a question could take the form of: 'by what year will there by 60% of households in the UK capable of receiving satellite TV with a 90% probability'? The responses are analysed to calculate the median and interquantile range of the dates. This information is recirculated, and those respondents falling outside the interquantile range are asked to give their reasons. The second-round responses are analysed; the respondents receive the new median and interquantile ranges together with the reasons put forward for the early or late dates of occurrence, and are asked to give revised forecasts in the light of the new information. This provides the data for the final forecast.

Scenarios are descriptions of alternative possible futures. The rationale is that for longer-term forecasts there are sufficient substantial uncertainties for several alternative futures to be equally feasible. They are often used where there are major social, political or economic uncertainties such as in the oil industry. Shell, for example, uses scenarios for its strategic and corporate planning throughout the company.

Relevance trees (see Chapter 4) explore the route by which a postulated objective might be reached, say the economic generation of electricity from nuclear fusion, and trace the technological and other developments which would be necessary for its achievement.

Chapter 4
Engineering, marketing and product development

The first three chapters have been concerned with the framework within which the engineer must work, the establishment of those corporate and technical areas where the main thrust of his activities must be directed, and the evaluation of the business potential presented by advancing technologies. Within this structure he must develop specific products which individuals are prepared to purchase. In the final analysis manufacturing companies do not make profits by having the right strategy; their future depends upon the design, manufacture and marketing of products which appeal to customers at a price they are willing to pay. Indeed some companies have prospered through an intuitive flair for the market without having an explicit strategy. Others with sound strategies have failed through their inability to translate them into the operational actions by which they can be fulfilled. This chapter will discuss the development of these products and the need for close links between the engineer and the market. In this chapter the word 'product' is used to embrace new processes and engineering projects as well as products marketed to the final consumer.

All too often new products are designed and launched into a non-existent market. There will always be a high failure rate because of genuine market uncertainties. However, research into many failures indicates that there was often never any likelihood of there being a market and the product was developed with inadequate or faulty market analysis. In order to eliminate these errors there must be a close relationship between the engineer and the market. Formally within the company this implies a close integration between the technical and the marketing departments. This is unlikely to occur unless the individuals in the two departments form a good working relationship and understand the other party's problems. For the engineer this means that he must always bear market considerations in mind when proposing new products. It also means that there is a need for the marketing man to have sympathy with the engineer and an understanding of his problems. Formal organisational structures and procedures help to bring this about, but more important are the attitudes of the individuals. For the engineer it means that he must play a role in encouraging the involvement of the marketing man and explaining the potential of his technology.

4.1 The new product concept

In Chapter 3 we discussed market pull and technology push. In the case of market pull the technologist is faced with responding to a market need which has been identified. The majority of these needs will involve incremental improvements in the performance of existing products, often associated with the aim of reducing their manufacturing cost. In general these will not lead to any significant change in the size of the total market. They are essential to maintain the competitive position of the business, with the hope that they may also yield an increase in the company's market share.

From time to time the marketing department will come up with ideas for entirely new products. All too often the initial reaction of the engineer is negative. This may be because he sees major problems of technical feasibility without perhaps devoting sufficient attention to how they might be overcome. In many cases the problems are genuinely insuperable and may confirm the engineer's opinions regarding the technical illiteracy of his non-technical colleagues. However, there are examples where organisational pressures have caused the technical department to initiate work against their better judgement which has eventuated in solutions leading to the development of successful products. Another barrier to the free flow of ideas is a reluctance to accept those originating from outside the engineering function – the 'not invented here' syndrome. It is important, therefore, that the engineer keeps an open mind to suggestions emanating from the marketing department, or indeed from any other source within the company. Resistance to these ideas may not only result in the loss of proposals which have merit but also damage personal relationships, thereby discouraging the flow of ideas in the future.

In the case of technology push we are concerned with the generation of new product ideas based upon the exploitation of a technological potential. But they are in themselves of little value unless they can be matched to a market need.

The influences leading to a product concept are shown in Fig. 4.1 and are discussed in the following sections.

Creativity
One can think of creativity as falling broadly into two categories – inspirational and creative problem-solving. The thought processes by which a new idea is born are imperfectly understood. The classical example of creative problem-solving is that of Archmedes jumping out of his bath shouting 'eureka'. The mind may be working subconsciously and the ideas may appear randomly. However, they frequently depend upon the association of two streams of thought, knowledge or experience which are not in close proximity when one is following a normal logical thought pattern. This seems to occur more frequently in some people – those we call creative – than in others. However, all people are creative to some extent. It is the quality of the ideas that ultimately determines engineering success.

The success of the engineering contribution to the business is highly dependent upon having a number of people who fall into the creative class and the reception they receive when they submit their proposals. These two aspects are clearly related. The key factor is the environment within the technical department, which is largely dependent upon the attitudes of senior management. If new ideas are not welcomed it will be difficult to recruit or retain creative people. Furthermore, there will be a reluctance for those who remain to submit proposals which require any significant change to the technical or managerial conventions within the organisation. Where senior technical management complain about the lack of ideas being presented to them, often implying that it is the fault of the training or the calibre of young engineers, the cause can usually

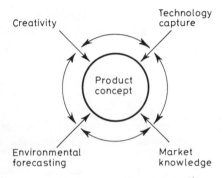

Fig. 4.1 *Development of the product concept.*

be traced to the attitudes of the managers themselves. The nature and quality of the senior technical managers are in turn frequently a reflection of the attitude of the business to innovation, risk and technology. It is important that the engineer understands the constraints imposed by the corporate environment within which he is working if he is to avoid frustration when his ideas are rejected.

Much of the work of the engineer is concerned with the solution of technical problems. In recent years a number of techniques have been developed for deriving creative solutions to these problems; they include brainstorming, synectics and lateral thinking. All these techniques are based upon freeing the mind from the constraints of normal logical thought patterns and suppressing premature evaluation, which may inhibit the full development of the creative idea.

In many companies the generation of ideas is left to chance; little effort is devoted to creating the conditions which are necessary to ensure a steady flow of new proposals. There is a tendency to adopt the first solution which presents itself – often a development of existing practice – rather than to seek actively for more original approaches. There is little the individual engineer can do to change the environment in which he works (though he should attempt to understand it), but he should pay attention to the development of his own creative ability and explore the possible contribution that can be made by the creative problem-solving techniques. He should also be aware that a high proportion of ideas will prove unacceptable on grounds of technical feasibility or because they might result in activities not consistent with the culture, strategies or policies of his company.

Technology capture
Even the largest companies with substantial R & D investment are only making a relatively small contribution to the international world of rapid technological progress. With the growth of data banks and data retrieval systems it is now possible to obtain information on a wide range of international developments. These formal systems can be extremely valuable in identifying sources of knowledge once the right questions are asked. The difficulty is in deciding what questions should be posed. In order to do this it is essential to maintain good contacts with the world outside the company in order to pick up early warning of significant advances. This is technology capture (see also Chapter 2).

The sources of information which could be of value in doing this are varied: technical journals, academic institutions, attendance at conferences, meetings of learned societies, the media and casual conversations. Much of this information, acquired randomly, is not of immediate use and will be stored in the minds of individual engineers. The richer the store the more likely is the emergence of ideas which could be fruitful. Systems for the scanning of technical literature are an aid but the value derived from them must depend upon the thought process they trigger in the individual. Nor should this be assumed to be a purely technical matter, since the ultimate object must be to link the technical with the market potentials. As we have seen earlier the feasibility of a product idea may depend upon development in several technologies or on market trends. These interactions must be identified in relation to the individual items of information and can be a valuable input to a monitoring system.

Research has shown the importance of *gatekeepers*. There are individuals with a wide range of contacts both personal and through the literature. They may be frequent attendees at conferences. As a consequence they hold a store of information, much of which may appear to be irrelevant at the time it is acquired but may prove to be invaluable at some future date. Such people emerge within an organisation and their value and activity is dependent upon their personal inclinations. They cannot be appointed; they need to be identified and used. The existence of technololgy gatekeepers has been proven in a number of research programmes. There are, however, gatekeepers in all functions. Marketing gatekeepers may be the first to see the changing needs of the consumer. Top management may be business gatekeepers in a broader sense.

Although this chapter is primarily concerned with the development of new products, the concepts of creativity and technology capture are also relevant to most aspects of an engineer's work at subsystem and component level. Thus he should take an active role in improving his own performance in these areas rather than rely solely on inputs from elsewhere in the organisation. He must develop his own means to capture and store key items of inforamtion, monitor developments and use the expertise of gatekeepers. In this way he will become better informed on a wide technological front and better able to make a contribution to the engineering excellence of his organisation.

Creativity and technology capture are closely related in that the former can be stimulated by the latter. Conversely a creative idea may not be exploitable without the use of a piece of technological knowledge which the company has not used before.

Market knowledge
All the arguments discussed in relation to creativity and knowledge capture are mirrored in the marketing department. There is need to identify novel consumer satisfactions which are not currently catered for. For example, the identification of a market for a coloured Wellington boot was a marketing stimulated act of creativity at a time when all the existing products were black.

Similarly it is necessary to acquire information on what trends in consumer tastes are emerging and on what competitors are currently doing or may be doing in the future. It is an important function of marketing to ensure that this information, even when it may be a half formulated idea, is communicated to the technical departments. For it is only the linking of an understanding of existing or potential consumer desires with the engineering knowledge base which can yield the first step in new product development – the product concept.

This process is unlikely to be effective in the absence of a genuine dialogue between the two areas of expertise. The essential first linkage must be between technical knowledge and consumer satisfactions at the conceptual level. Only when this has been achieved should attention be devoted to the physical nature of the product. This dialogue is often lacking. Formal methods of communication may assist but are unlikely to achieve a great deal unless there is mutual respect and understanding between those who understand the technology and those who understand the market. Furthermore, as has been stressed previously, this is unlikely to occur without the active participation of the engineer. In its absence good market concepts will be rejected after a superficial technical examination, or products will be developed for which there will never be a market. These points, although they may seem obvious, cannot be overemphasised, since research into product failures indicates their significance even in large companies with sophisticated formal systems.

It is often helpful to analyse the attributes of a technology and to attempt to link them with the attributes which consumers may wish to have satisfied. For example, the attributes of a ground effect vehicle are: low friction, load spreading, and lack of physical contact. One or more of these contributed to the success of such products as the Flymo lawnmower (low friction), vehicles for transporting heavy loads (load spreading) and hospital beds for severely burned patients (no physical contact). In retrospect it is easy to identify the linking of the two sets of attributes on which depended the success of the products designed to exploit them. The identification of these linkages may be revealed by systematic analysis but is more likely to emerge from the exercise of creativity. But an understanding of the attributes assists in the creative process.

Another important source of ideas is the user himself. This is particularly valuable in process industries and industrial market where a sophisticated user is likely to incorporate modifications of his own to improve the effectiveness of his equipment. Close contact with the user will reveal these improvements, leading to their incorporation in existing designs or an entirely new product. Normally information on these modifications will not be provided by the user; the manufacturer must find out for himself. Although the normal channel of communication is through the marketing department, this should not prevent the engineer from having access to the customer. All too often the engineer remains closeted within his own area of operation, particularly in large companies, and loses contact with the real world where his efforts have their impact.

In some cases the engineer should gain experience as the user of his own products. Aircraft designers cannot be expected to be pilots but it is not unrealistic to expect a designer of tractors to learn to drive the product. In one large tractor manufacturer it was found that none of the design staff was permitted to drive the product; it is not surprising that in these circumstances there were aspects of the tractor which were not 'user friendly'.

The conclusion must be that the engineer, whilst working closely with the marketing department, should also try to identify with the needs of the customer. This will almost always lead to better products and occasionally to the generation of ideas for entirely new products. For example, an analysis for the engineering configuration of existing products may reveal that there is a major component which performs a function required of products to satisfy a wide range of user needs. This can be seen in the domestic equipment market where a wide range of products are based upon the electric motor. Thus Black and Decker developed a wide range of products utilising their electric motor technology, whereas potential competitors remained wedded to a limited range of traditional products (for example the vacuum cleaner) with which they identified closely. A simple shift in orientation from 'we are a vacuum cleaner company' to 'we are a company based upon electric motors' opens up the possibility of significant market diversification. Although this is a corporate decision the stimulus for it comes from a realisation that the technical strengths (e.g. electric motors) can be matched to a marketing strength (e.g. experience of the domestic equipment market) to create new opportunities. In retrospect the example given above may seem obvious but it was an opportunity not seized by many leading companies.

Environmental forecasts
Technology forecasts have been discussed at length in Chapter 3. Their importance in relation to developing a product concept is that they indicate what will

be feasible in the future. Particularly where development times are lengthy it is important to assess how future advances might yield an opportunity for new products which are not currently feasible. In such cases it is unlikely that product development could be initiated immediately, but the forecast can guide the programme of exploratory research to ensure that the full potential is exploited at an early date.

Economic, social and political forecasting have a more direct impact on market factors – what the consumers may require, and their ability to pay for it.

The evolution of the product concept
At the time when a product concept is generated, ideas for the physical form of the product itself may be poorly defined. Time spent in generating the concept may appear frustrating to the engineer impatient to get to grips with the technical problems of product development. However, it is an essential first step which cannot be rushed. Rarely will engineering excellence overcome a failing in the original concept. It is also the highest level at which market and technical potentials are married.

Although the four elements in concept generation – creativity, technology capture, the market and environmental forecasts – have been discussed separately, they are interactive. The initial stimulus may come from any of these sources, but the processes involved are iterative and require the interplay of knowledge and insights possessed by different people in the organisation – but most importantly engineers and marketers.

4.2 Product specification

The product concept stage will have established the potential of a new or existing technology or combination of technologies to satisfy the needs of a new or existing market. The next step is to draw up a specification for a product or range of products which will capture all or a share of these markets – what it should be and who will buy it. The specification itself must consist of three factors:

Performance This relates to the physical performance characteristics of the product itself – what it will do. This is frequently referred to as the technical specification.
Time The date at which the product will be ready for market launch.
Cost which consists of two elements: (a) the unit cost of the product; and (b) the total development costs, which represent the R & D, design, production equipment investment, and marketing costs to the time of launch.

It must be appreciated that there is a wide range of combinations of these factors leading to different technical configurations. There will normally be trade-offs between the factors, for it is highly unlikely that the product with the highest performance will be the quickest and cheapest to develop. Thus a conscious choice has to be made as to the optimum combination.

The choice must be made in relation to market and financial criteria. All things being equal, one would expect a higher performance to appeal to a larger market. In practice, however, the reverse is likely to be the case since the development and unit costs are also likely to be higher. Hence it is necessary to relate the technical alternatives to a chosen market segment within the overall corporate and technological strategies. Competitive products, present and future, must also be considered. A short development time may justify the additional development cost if it enables the company to launch its product first; it may also justify a lower initial performance provided plans are made to upgrade it later in order to maintain the competitive advantage.

It can be seen that the technical and market considerations and alternatives must be examined together. In doing this the decision must be market driven in that the product design must be matched to a specific group of potential customers after taking account of the technical specification constraints. Once this matching has been achieved there is little room for technical manoeuvre. Frequently there is an inclination to give the market more than it asked for. This over-design will inevitably incur cost and time penalties, with implications for unit cost and product launch date which detract from the market success. A justification for over-design might be the potential it yields for the future expansion of the market through subsequent products with an enhanced performance; in such cases this should be undertaken as part of a comprehensive marketing plan and not decided unilaterally by the engineering departments.

The business purpose of a new product is primarily its profit contribution in relation to the development costs. Although this is a factor that the marketing department will have in mind, they often place more emphasis on the absolute size of markets. A high-margin small market segment may well contribute a greater profit than a low-margin mass market. Such considerations have an impact on the engineering specification but also on the ability to satisfy it. If the analysis indicates a radical change from previous practice, major difficulties may be experienced. In several instances aerospace companies accustomed to designing to high engineering standards have found it impossible to drop these standards sufficiently to meet the low costs demanded when the company has attempted to enter consumer markets. The opposite effect has also been noted where companies have attempted to move up market.

Manufacturing investment must also be considered. The size of the market segment will determine the appropriate manufacturing technology and the associated capital investment. Since the product design must be matched to the method of manufacture, further technical implications will arise.

Thus it can be seen that the setting of a product specification is highly complex, involving a large number of technical, marketing, financial and manufacturing considerations. These decisions cannot be taken in isolation since the aim must be to optimise the trade-offs in such a way as to maximise the return on the development costs, primarily in the short term but also with an eye on possible future improvements. The decision will depend upon the interaction of many individual judgements. It is essential, however, that the agreed specifications be adhered to and only modified in the light of new knowledge and a complete review of its total business implications.

Product specification and the industry life cycle
The characteristics of the industry/technology life cycle described in Chapter 2 have an influence on the emphasis to be placed on the trade-offs between performance, cost and time in the specification. In the early stages of the industry (phase A in Fig. 2.3), development performance and the date at which it is achieved are paramount. This can be costly, since if the industry is to be successful the cost considerations must be overruled in the interested of speedy development. Attention focused on cost minimisation will almost inevitably result in the loss of competitive advantage, associated with low sales revenues and often resulting in financial loss. Of course, needless waste must always be avoided, but to be successful the money must be made available to sustain the maximum rate of development progress. If the financial resources to do this cannot be found then the development should not be initiated. Furthermore, in assessing the financial requirements the need to fund further developments to maintain a flow of products with increasing performance must not be overlooked if an initial competitive edge is not to be surrendered.

In phase B of technological growth the need for urgency diminishes. More important is the identification of market segments. Development cost becomes a more significant factor. By this time the main features of technological or product development will be established. There will be a wide range of possibilities for designing a selected range of products catering for specific market segments, only some of which the company may wish to enter. The key decision is the choice and the appropriate trade-offs between performance, time and cost.

In phase C, the mature phase, time becomes of decreasing importance owing to the low rate of industry growth. The performance improvements to gain a competitive advantage may be quite small. However, as we have seen earlier, technical (R & D) productivity is now low and large investments may be necessary to achieve a marginal improvement. Consequently the cost element of the specification will dominate.

Attribute analysis and market segmentation
At the product concept stage technical and market attributes were considered primarily in relation to the generation of concepts for entirely new products. These may be the most important for the long-term future of the company, but the greatest number of new products are launched in phase B of the industry growth cycle. The market is now established and the proliferation of alternatives arises from the design of products to satisfy the growing emphasis that cus-

tomers place on their own individual needs. In the days of Henry Ford's black Model T, customers were satisfied with a motor car which would meet their need for physical movement. In today's market attention has moved to a wide range of other attributes; these include comfort, speed, reliability, durability, aesthetic appeal, price, running costs and ease of maintenance.

In the limit each customer would like his own bespoke car, reflecting the weightings he attached to each of these attributes. The aim of market segmentation is to identify a group of customers and estimate the number who could be satisfied by a particular combination of these individual weightings. When this has been done for a chosen segment it can be translated into a series of technical requirements, e.g. road holding, speed, acceleration, servicing cost and interval, and corrosion resistance. The aim must be to achieve the minimum performance necessary to satisfy the needs of the chosen segment. Any improvement on this level will almost inevitably raise the production cost and thereby the price, a factor which is weighted highly by most potential purchasers. The idea of designing down to a minimum acceptable level does not appeal to many engineers inculcated with the desire to achieve technical excellence. However, if these ideals prevail the inevitable result is to ask the customer to pay for something he does not really desire, with the associated danger of pricing the product out of the market.

It might appear from the discussion so far that the marketing department should dictate to the engineering department. But this is not the way in which the specification should be determined. Collaboration is essential at the various stages in the formulation of the specification since many elements of it are difficult to quantify at the outset. Market research can be expected to provide guidance on customer preferences, but it is not always easy to establish relative weightings. Another problem is the translation of some of the attributes into terms which have a meaning for the engineer. Motor car comfort, for example, consists of several features which can be considered separately such as seating, environmental control, road holding and quietness. Some of these will lend themselves to quantitative expression; others will not. Where the requirements cannot be reduced to quantitative technical specifications the engineer must try to understand what the marketing man requires, and the latter must appreciate the technical and financial implications. Thus we see here (as in so many other areas) that the needs of the organisation can only be achieved through dialogue, understanding and in some cases negotiation between individuals with different functional orientations, values and attitudes. This can only be achieved by a combination of both formal and informal communication.

Fig. 4.2 shows diagrammatically the process described above. The end product of this process must be a firm specification in relation to performance, cost and time which provides the basis for the establishment of engineering objectives and plans. There may, of course, be a need to incorporate modifications to the plan at later stages owing to problems or to new knowledge arising from the technical work or the market place.

Engineering, marketing and product development 89

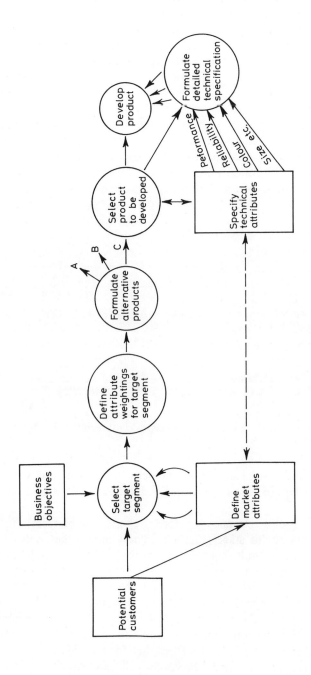

Fig. 4.2 *Matching market attributes to the technical specification.*

4.3 Exploratory research and development

In the early stages of concept generation and project definition it may not be certain that the technology is capable of achieving the desired ends. These uncertainties can manifest themselves in two major ways:

1 Lack of theoretical knowledge
2 Doubts about the ability to apply the knowledge, and if so the financial implications.

In general industrial companies do not undertake basic research. On occasions, however, it may be necessary to do research to establish the feasibility of satisfying a product concept. For example, it was necessary to conduct research into the problems of re-entry into the earth's atmosphere before the first ballistic

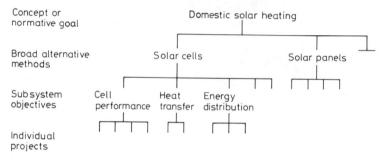

Fig. 4.3 *The relevance or objective tree.*

missiles could be designed; several alternatives were explored. Another example is the research into the properties of ceramic materials prior to the design of a ceramic internal combustion engine. Where the uncertainties are great it will normally be necessary to complete the research before the product specification. Where there is a greater degree of confidence that a solution will be found the research may be carried out in parallel with the early stages of product development.

The drawing of a relevance (or objective) tree can be helpful in structuring one's thoughts at this stage (Fig. 4.3). The tree is a hierarchy of objectives, starting with the overall objective of the project or the product concept and setting objectives for the major technical functions and systems to explore feasibility and alternatives. At each stage the tree permits the exploration of either alternative approaches to meeting the sub-objective or the constituent subsystems. Each alternative can be analysed in relation to performance, time and cost. Thus a development programme to minimise cost might indicate a different combination of projects from that for minimum time.

In particular the relevance tree approach can be extremely useful in the following ways:

Engineering, marketing and product development

1 To establish whether a specific normative goal or objective *might* be achieved. Initially a concept might be merely a desirable target without there being any clear ideas of how it might be achieved in practice or whether indeed it is feasible with current or foreseeable levels of technological performance.
2 To determine alternative ways by which the objective may be met. There may be a number of alternative routes to the desired end point. This structuring of the problem can aid the creative process.
3 To identify major technical uncertainties. This could indicate the need for exploratory R & D or detailed technology forecasts.
4 To decide performance objectives for the constituent parts of the programme in order to achieve the overall objective.
5 To select the optimum programme of work and individual R & D projects from the alternatives already identified.
6 To assist in the communication of the above.

Even where the theoretical knowledge exists there may be doubts as to whether it can be made to work in practice, or indeed whether the technology exists to manufacture it. In this case applied research is required to establish this

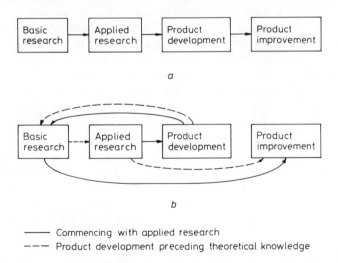

Fig. 4.4 *Models for technical development: (a) the logical sequence, (b) alternative models found in practice.*

feasibility. For example, in aerospace theoretical work may suggest advantages in a particular wing configuration. This would normally be followed by two stages of applied research, firstly in a wind tunnel and then on an experimental flying model. Only on the conclusion of these programmes can development of a saleable product be initiated.

Fig. 4.4a shows this sequence of events. The stage by stage process follows a logic with which all engineers will be familiar. This minimises risks in that all major technical uncertainties are removed before moving from one stage to the next. However, when one examines a number of successful innovations it is apparent that this logic has not always been followed. In its place the much riskier process of moving straight from concept to development has been adopted on the assumption that any problems which arise can be solved as they occur (Fig. 4.4b) The development of the float glass process by Pilkington is an outstanding example of the successful employment of this approach. Without the theoretical knowledge many major and costly problems were encountered; the theory came later to explain why it worked and to enable impovements to be made. If the project had failed it would have been difficult to defend the actions taken. But it did not fail and, in spite of the problems encountered, the rewards from completing the programme before finishing the research more than outweighed the additional costs incurred.

Thus we have two alternative models. The first minimises technical risk and business risk but may reduce the potential benefits. The second ignores the technical risk and maximises the business risk but also offers the opportunity for greater profit if successful. The attitudes of engineers to the second approach will be influenced by the source of the impetus for develoment. Few objections may be raised if the orignator is technical, as was the case with float glass. However, if the demand comes from elsewhere in the organisation the engineers will probably object strongly to initiating a programme where they have no theoretical grounds for confidence in success. If the potential rewards from success are substantial then the associated risk may be justifiable. This is a business risk and it must be assessed in relation to corporate aims.

4.4 Project evaluation

The establishment of the feasibility of a new product is not sufficient grounds for going ahead with the development. In a dynamic, creative company there will be more potential ideas than the company can support. A selection has to be made. Even where only one project is being considered it must still be assessed.

The extent of the evaluation process at the outset will depend to a large extent on the nature of the product, closely related to its position on the industry life cycle. For a radical new product there are two questions which must be addressed:

1 Will it satisfy a need that cannot be met with existing technology; or
2 Will it satisfy a need currently being met by existing technology but at a considerably lower price?

However detailed and professional the development of a radical new product may be, there are always poblems encountered by the first users. So we can assume that no rational person would ever purchase a product incorporating new technology if he could avoid it. If the answers to the two questions are 'no' then there is little prospect of the product succeeding. If the answer to one of them is 'yes' there will be grounds for proceeding, either because the performance is such that the customer is prepared to accept a degree of inconvenience in order to meet his needs, or because the financial benefits outweigh the disadvantage.

In a mature industry a detailed evaluation can and must be undertaken before development commences. Because the company has data from the past and the level of innovation is low, reasonably good estimates can be made for development cost, product price and development time. The market analysis should also give a reasonably accurate estimate of the market size and expected sales. These can be combined to yield the expected profit level and return on the investment. However, small errors in these markets can have a disproportionate effect on profitability since the margins can be expected to be low in what is a highly competitive market. Thus the success of the product is highly vulnerable to small errors, changes in the economic environment and the actions of competitors. A minor technical fault, for example initial unreliability, can have a major impact on market acceptance and the business as a whole.

Between these two extremes lies a range of engineering industries where there is an element of technical risk (but less than with the radical innovation), some uncertainty with the size of the potential market, and competition for funds from more projects than the company can support.

Thus it can be seen that the project evaluation system may need to be tailored to the type of industry or product. In all cases, however, the proposal must be evaluated in relation to its total business impact, namely:

Corporate objectives, strategy and values
Technical
Marketing
Manufacture
Financial
Procurement.

Under each of these headings there will be a number of criteria which must be evaluated. These will vary with the industry, but the list must be comprehensive to ensure that no factor relevant to the success of the product will be overlooked. Some items such as market size, market share, product life, development cost and product cost will be common to all evaluations.

The difficulties arise when the individual project proposal is assessed against this list of criteria, since reasonable estimates may only be possible once detailed design has been undertaken. It is easy to stipulate a product cost in a specification; it is far more difficult to design the product to meet this cost. Thus the first time the evaluation is done some major unknowns are likely to remain, particularly in relation to engineering, marketing and manufacture. Nevertheless the evaluation may reveal enough information to reject the proposal because it fails to satisfy one or more of the criteria. For proposals that appear promising, approval to proceed with the development should not be taken to guarantee that it will necessarily be allowed to continue to completion. Initial views of the prospects for a new product are usually optimistic, particularly in relation to financial estimates (see Chapter 6); thus proposals which appear marginal must be regarded critically.

In many organisations project evaluation is seen as a technical function. This is wrong. The development of a new product is a business undertaking, not a technical project. Not only should non-technical criteria be considered, but the involvement of those who can make the individual judgements – corporate, marketing, financial and manufacturing – is essential. The uncertainties which must be resolved as the project progresses fall into all these areas. Periodic re-evaluations are also essential to ensure that new information bearing on the success is incorporated in the overall assessment of whether to continue or to terminate the project.

The main purposes of an evaluation system are:

1 To ensure that no factor relevant to the commercial success of the project is overlooked
2 To reject those proposals which fail to meet the criteria
3 To record the assumptions and estimates made at the time the project is initiated
4 To identify major areas of uncertainty – technical, marketing, financial and production – as a basis for further investigation
5 To provide a common basis for evaluation when choosing between competitive proposals
6 To ensure the involvement of non-technical functions in the decision-making process.

A considerable number of formal evaluation techniques are described in the literature, ranging from simple check lists to sophisticated mathematical models. Their suitability will depend upon the characteristics of the industry and their acceptability to those involved. A comprehensive check list of criteria is essential. It may be questioned whether the more elaborate techniques are useful in view of the uncertainties; indeed, they may be counter-productive in that they can give a spurious impression of accuracy.

Another shortcoming of many of the techniques is that they can favour mediocrity. Most successful products exhibit one outstanding feature; this may be technical performance or reliability, low cost, or market appeal. This one feature may be of greater significance in making the buying choice than other deficiencies, provided they are not too serious.

Projects do not succeed solely because they satisfy a set of criteria. This is no guarantee that they will generate the enthusiasm necessary for taking it to a successful conclusion. The importance of the project champion and people dedicated to the success of a project is a critical feature which will be discussed later.

The need for periodic re-evaluations was mentioned earlier. As a project progresses so will the uncertainties diminish. New judgements will be made for many of the criteria. Comparing these with those made at the time the project was initiated will reveal any market changes. This is an important reason for recording the assumptions and estimates made at the outset, since the comparison provides a measure of the impact of the new knowledge acquired upon the anticipated success of the product. The role of the evaluation procedure is therefore comparable with that of a control system whereby new information is compared with the old, the change of variance is noted and the appropriate action is taken. This action may be the cancellation of the project. More often it will require an amendment to the specification in the light of technical problems encountered or changed market conditions. The latter can be the most important. Frequently these changes occur slowly and their impact on the product design may be overlooked in the absence of a formal review where all aspects are re-examined. Where product development times are long it is not unknown for the engineering department to meet the original specification on time only to find that the product is no longer relevant to the market. This could not occur if the engineering and marketing departments worked in collaboration throughout. The evaluation process is a mechanism for improving this collaboration.

4.5 Product design

Once the specification is agreed the engineering design may proceed. If the procedure outlined above has been followed one would expect that the design to meet the specification would incorporate all those features required by the market. This may not necessarily be the case. The engineer through his training and inclinations will be motivated primarily by his desire to meet the functional performance requirements in the narrowest of senses. He may lose sight of some of the market attributes not related to the functioning of the product. Aesthetic appeal, for example, is important in all products; it is often essential for consumer products, and may also be a selling factor for many industrial products.

The word 'design' is used in a variety of senses. To many people it is associated solely with aesthetics, and is often regarded as a cosmetic exercise to be carried out at a late stage in the development process. Rarely is this adequate; it needs to be incorporated from the outset. In the wider sense in which the word is used today it refers to the identification of all the user needs, the required attributes, and their satisfaction by the final product. This emphasis on product design is to a large extent a reflection of the absence of the approach in many of the products offered to the market in the past.

Design for manufacture
The design process is often thought of as sequential. First the product is designed; then consideration is given to how it should be made. In many companies there is a production engineering department responsible for the design of jigs and fixtures and the manufacturing processes. Where no consideration is given to the methods of manufacture, designs may be produced which are impossible or extremely expensive to manufacture. This is highly undesirable, and manufacturing should be involved with the design process from the outset. This can influence not only the physical aspect of the design but also the choice of materials. We see here once again the need for an integrated business approach to the job of the engineer.

It will be seen in Chapter 5 that with the introduction of the microelectronic-based technique for design and manufacture a more integrated approach is being forced upon many companies. Not only does the absence of integration lead to cost escalation; it also introduces delays which may have a major impact on the profitability of a product with a short market life. This is of critical importance for those industries, such as electronics, where market lives can be of short duration. But in all industries development time is a significant determinant of market success. It has been suggested earlier that it may be wise to incur additional cost to reduce this time; however, good planning and the integration of activities can make a significant contribution, often saving expense as well as time.

Manufacturing equipment design has progressed through three stages, although all three still exist. In the first stage general purpose machines were the norm so that a change in product design had little effect upon the capital investment in manufacture. With the growth of automation, specialised equipment designed to meet the needs of a specific design became commonplace; when the product changed the factory was re-equipped with new machines. The costs of doing this increased significantly with the growth of sophistication; fortunately it is now possible to design for greater flexibility, again largely due to the application of microelectronics.

In many industries, however, the design of the manufacturing equipment is still inflexible. In one company with a newly installed manufacturing line it was necessary to spend over £1 million to convert it to producing a simple product in metric sizes. It is, of course, not always possible to foresee what changes will be required in the future. Forecasting and a manufacturing strategy can eliminate some of these problems, but in a rapidly changing world they cannot ensure that all future circumstances can be foreseen. The pace of change is accelerating in most industries. Thus it is highly desirable that the designs of both the product and the manufacturing processes are sufficiently flexible to enable these changes to be incorporated. Unfortunately flexibility usually involves a cost penalty which will affect the profitability of the initial product. Thus one is incurring a cost in order to reduce the possibility of a future risk. This must be a business decision, not one taken unilaterally by engineers. However, it is the engineer who can appreciate the technical and cost implications of building flexibility into his design. He must advise on these alternatives, but the final decision must be taken at the corporate level.

The economic manufacturing process is frequently dependent upon the planned volume of output. The unit cost of production consists of two parts: the fixed cost, largely dependent upon the sophistication of the equipment; and the operating cost, which decreases with an increase of manufacturing investment. For a given production volume there will be an optimum between these two elements which yields the lowest unit cost. To a great extent this can be seen as a consequence of the decision on the target market segment. There are, however, two considerations which should not be overlooked. Firstly, with a higher volume the magnitude of the reduced unit cost might influence the choice of market segment. Secondly, the initial volume may be expected to be lower than that achieved when the new product has gained market share. There are two alternatives:

1 To plan manufacturing on the basis of the final volume, accepting higher initial unit costs due to low utilisation of the equipment.
2 To accept that it will be necessary to change the manufacturing equipment at a later date. In this case the product must be designed to minimise the need for modification to suit it for the new manufacturing processes.

Design for minimum cost
Every engineering decision has a cost implication. Throughout the design phase the engineer must seek alternative ways to achieve the same functional performance at a lower cost. This can be obtained by the use of non-traditional technologies, different materials, and cost conscious detail design. It follows from the previous section that the relevant costs must include the cost of manufacture, implying that the design engineer should posses some knowledge of production engineering or work closely with those who do possess it.

In the final analysis it is not only unit price which is of interest to the customer. He is concerned with the total costs he incurs throughout the lifetime of the product. These consist of:

Capital cost This is the price he must pay to buy the product.
Running cost This depends upon:
1 The performance of the product, e.g. energy costs.
2 The reliability, often measured as the mean time between failures (MTBF). The criteria will vary with the nature of the product. Where safety is an important factor, e.g. aerospace, or where the failure of one part of the system has system-wide ramifications, e.g. oil rigs, reliability may be the major attribute required. Where replacement is cheap, e.g. electric light bulbs, price may be much more significant. The design must reflect the appropriate needs of the customer; a design which incorporates an MTBF of 25 years is likely to incur a cost penalty that the customer is not prepared to accept.

98 Engineering, marketing and product development

3 The maintenance needs, which can be a major factor in some buying decisions owing not only to the cost but also to the penalty of inconvenience (difficult to quantify). The servicing interval is an important attribute for the motor car.

All those considerations must be borne in mind during the design phase. In many industries the techniques of value engineering or value analysis are applied. Valuable as these approaches are it must be recognised that they are an admission of failures in the original design, which should be minimised if the engineers responsible were fully appreciative of the cost implications of their designs.

Design and the operating system
Many engineering products are embodied in or form part of a total system involving a number of manufacturing companies. Problems arise at two levels:

1 The overall technical system
2 Programme co-ordination.

In theory a full specification should ensure that the product is wholly compatible with the system for which it is designed. For this to be fully effective the design must be 'frozen' at an early stage. This is rarely possible, and many changes and modifications are made by the companies responsible for parts of the system as the design evolves. Frequently this gives rise to incompatibilities which only become evident when the total system is finally assembled; the classic complaint is 'It doesn't fit'. Formal systems for the approval of modifications are essential. In spite of this some errors can slip through the net if there is not a close integration between the contractors. The individual engineer should not shelter within the formal procedures; he should make every effort to ensure that he is, at all times, familiar with the design of the other components of the system with which his interfaces.

Many engineering programmes are delayed through the late delivery of a vital part of the system. Although there will always be some form of co-ordinating authority it may have little direct influence on the activities of the individual supplier. If that supplier is late in delivery the financial consequences for the total programme are likely to far exceed any possible penalties which can be imposed. The short-term business risk to that supplier can be minimal but the long-term consequence can be the loss of future contracts.

In an increasing number of industries the system is a combination of hardware and software. The remarks above apply equally in this situation since in many cases the software supplier is not the hardware manufacturer.

Design for market diversity
In some industries, notably electronics, the manufacturer is catering to a number of distinct markets – consumers, other manufacturers, financial services and

retailers. Each of these has its own characteristics and emphasis on the individual market attributes. Initially these distinctions may be unclear, not only to the design engineer but also to the marketers. The result can be a product which fails in one market whilst being a success in another area which dominated the thinking at the design stage.

This is a form of business diversification based upon the established technology. The aim must be to exploit one basic design as far as possible but to ensure

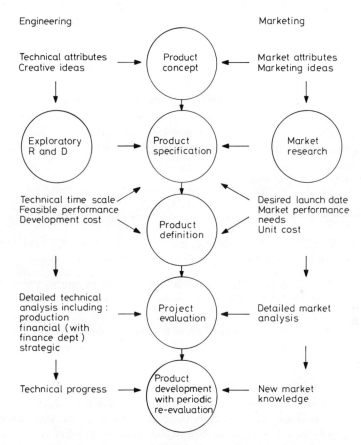

Fig. 4.5 *Integration of engineering with marketing at all stages of development.*

that it is capable of satisfying, perhaps by modification, the needs of the individual market. Many readers might consider this role to be the responsibility of marketing. Unless the marketing department employs people with a technical background, often the case in industrial markets, it is unlikely that they will identify the engineering implications of what may appear minor differences. The design engineer should be prepared to visit potential customers to find out for himself what they really require.

4.6 Engineering and marketing integration

In the preceding sections it has been stressed that engineering and the market must be matched at all stages from the original concept of the product to its detailed design (see Fig. 4.5). This is true of all products, be they the result of technology push or market pull. Companies can be expected to have formal systems for feeding information between the two functions. These procedures should ensure that both engineering and marketing requirements are co-ordinated. The procedures, although providing an essential framework, are insufficient by themselves. True co-ordination can only occur at the level of the individuals. In this the engineer should take an active role, not only through the marketing department but also through direct contact with the market itself. Some companies discourage this direct relationship, regarding it as a usurpation of the marketing department's responsibility. It should not be so regarded. However, it should only be done with the knowledge and co-operation of the marketing department, who should normally be involved in arranging the meetings and accompanying the engineer on his visit. Each must know what the other is doing.

4.7 Cost and pricing

The criteria by which products and engineering projects are judged are discussed in Chapter 6; the most important of these are:

Profit which is a measure of the total financial contribution to the company
Profitability which relates the profit earned to the investment required to achieve it
Cash flow which is the rate of cash drain on the company's financial resources during development and is the rate of inflow of cash when the product is earning.

Profitability is the main concern when evaluating projects. However, it is the maximisation of the absolute level of profit from all activities which must be the company's ultimate aim. In most cases these two criteria do not lead to different choices, but this is not always so. Sometimes the profitability of a few large projects may be lower than that of a number of small projects. Because of their small size the latter will each contribute less profit in absolute terms, although the aggregate of their profit contribution will be greater. However, this advantage may be outweighed by the problems associated with the complexity of managing a large portfolio of products or projects. Thus, although financial evaluation is normally carried out initially in relation to the individual project, the business concerns must be focused on the portfolio as a whole. In some cases this may lead to the exclusion of a proposal which appears attractive on its own merits.

The cumulative cash flow can be regarded as a constraint. Where there are only a few engineering projects at any time the cash drain may be greater than the financial resources of the company can support, however attractive the long-term profitability may appear.

Within the framework discussed above, choices have to be made of the form of the engineering design in relation to the anticipated market sales volume. The first point which must be stressed is that the selling price of the product must be established in relation to what the customer is prepared to pay for it. This is quite distinct from the cost the company incurs in making the product. In many engineering companies it has been traditional to price on a *cost-plus* basis: the cost of manufacture is determined and a profit margin is then added to establish the price. This was common in many government projects. It is highly undesirable because:

1 The manufacturing cost is of no interest to the customer and does not relate to the price he is prepared to pay.
2 The business aim must be to maximise the profit contribution. The profit is a residual determined by: sale price less manufacturing cost, multiplied by sales volume. It will be seen later that sales volume is dependent upon sales price and that manufacturing cost is related to sales volume.
3 It leads to engineering inefficiency and poor decision-making. If a predetermined profit margin is added to the cost computation there is no incentive to reduce cost; indeed, the higher the cost the higher the profit. The consequence may be to price the product completely out of the market or severely reduce the sales volume potential. In the long term the future of the company is prejudiced by the encouragement of poor engineering practices.

Although cost-plus is rarely used today as a basis for negotiating engineering contracts, much of the thinking it engendered still persists. The engineer must appreciate that cost and price are independent. This does not mean, of course, that cost estimates can be ignored, for they are a major determinant in assessing the likely profit margin and the viability of the product.

Price–volume relationships
Most markets have a *price elasticity*. This means that the lower the price the higher the sales volume is likely to be. In recent years the importance of this relationship has been decreasing in many markets. This is a reflection of growing consumer affluence, resulting in a relative downgrading in the weighting of price in relation to the other product attributes. The consequences of this for the engineer is that his designs have to reflect this change of consumer preferences. Thus although the cost of a given design must be minimised, the design itself should not be 'designed down' to a low price by the sacrifice of features desired by the market. A manifestation of this trend is the importance today attached to the wider aspects of design. Notwithstanding this, the price–volume relationship must still be considered.

This has an impact on engineering in two ways:

1 The higher the volume, the greater is the advantage obtainable from investment in manufacturing processes, thereby raising the fixed element of manufacturing cost (i.e. equipment) and reducing the variable element (i.e. labour).
2 The product design must be related to the type of manufacturing process.

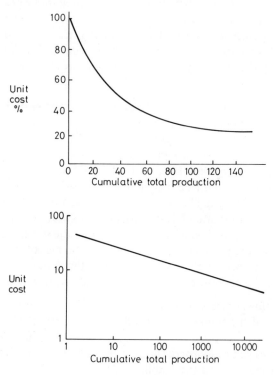

Fig. 4.6 *The experience (or learning) curve: (a) normal scale, (b) logarithmic scale.*

Thus it can be seen that the selection of price cannot be made in isolation from a consideration of the expected sales volume, the choice of manufacturing process and the design of the product. Involvement of all three functions – marketing, design and manufacture – is essential.

Unit cost and cumulative production (the experience curve)
Experience with the manufacture of a product leads to the incorporation of various improvements in design, manufacture and management. Their combined effect is to reduce the unit cost of manufacture through what is termed the experience or learning curve effect (see also Chapter 6).

When unit cost is plotted against cumulative production volume a curve of the shape shown in Fig. 4.6a is found. When plotted on logarithmic graph paper this takes the form of a straight line with a slope normally between 70% and 80% (Fig. 4.6b). This means that every time the cumulative volume doubles the unit price falls to 70% of what it was before. The relationship holds for a wide range of products internationally. It is a vital piece of information for product/market planning, and all companies should be aware of it. The actual slope will vary with the characteristics of the technology and the product as well as the effectiveness of the management. Examination of a typical curve reveals a number of factors which must be considered both in pricing the product and in its design, namely:

1. The initial unit cost of a new product will be high compared with what will be achievable later.
2. The new product can only be sold profitably if
   (a) The price is high, implying that its performance advantage is such that customers will be prepared to pay the high price. This will usually be a specialist market where the sales volumes are likely to be low. *Or*
   (b) The unit cost is significantly lower, so that the launch price can be comparable with the competing product which has already benefited from the experience curve effect.
3. The large-volume producer will be more profitable than the small producer where the products are similar, since the prices charged by both cannot be significantly different. This suggests that market share is a major determinant of profitability, and indicates that the small company is more likely to succeed by adopting a strategy whereby it seeks relatively small specialist markets where it can capture a high market share rather than compete with the market leaders in the high-volume markets.

The experience curve effect can be a great inhibitor of engineering innovation in a mature industry, particularly if it is not possible to identify unsatisfied niche markets. For in these industries radical innovation which would enable a significant improvement in unit cost is unlikely. Indeed the initial unit cost of the new product may well be higher than that for the existing product. Direct comparison of these two costs without appreciating the cost improvements which will occur as the cumulative volume increases may lead to the rejection of the new product even where its future profit potential is attractive. Comparisons must be made with the situation when the new product has established itself it they are to be realistic.

Thus in order to establish itself it may be necessary to launch the new product at a price lower than its cost, this is sometimes referred to as 'experience curve pricing'. The aim in doing this is to build up volume quickly to achieve break-even at an early date followed by increasing profitability as the real cost benefits from the new product manifest themselves. Such a policy is only feasible if the company has the financial resources to support this initial negative cash flow.

Consideration of the experience curve introduces an additional complexity into the new product development decision, since the dynamics of the market and the cost–volume relationship are varying with time, or more correctly with the cumulative volume. This has significant implications both for engineering design and manufacturing investment. The balance between the long and the short term must be considered, as also the element of risk. If there were no uncertainties regarding the long-term success of the product the short-term cost and cash flow penalties would normally be acceptable in the anticipation of higher long-term profit margins. However, uncertainties do exist and the choice of design and manufacturing process will depend upon the assessment of the magnitude of the risk and the corporate attitude to risk. Alternatives must be explored and it must be recognised that the choice is a corporate not an engineering decision.

4.8 Product development

The stages of product development are discussed in relation to their financial implications in Chapter 6 (Fig. 6.2), where it will be seen that the rate of investment in a project increases as it advances through feasibility, prototype construction and testing, and development of the product to be marketed. The aim is to recoup this investment as quickly as possible, and the minimum cost programme is unlikely to be the optimum in relation to the organisation's overall financial objectives.

The aim in development must be to ensure that the new product reaches the market at the earliest possible date. There are a number of reasons for this:

(a) The launch of a new product before that of a competitor gives the company the marketing initiative and the opportunity for establishing a dominant position.
(b) Where the product is protected by patent it extends the duration of its market life before the patent expires.
(c) The beneficial impact on cash flow.

There are five ways in which development times can be reduced:

1 Good project planning, control and management
2 The concentration of resources on a few projects
3 The adoption of CAD and other IT-based systems
4 Additional financial investment in the project
5 Acceptance of higher risk.

It is worth noting that the system of project management described in Chapter 5 can make a valuable contribution to reducing development times by improving functional integration. This removes many of the problems associated with the traditional sequence of limited co-ordination between departments. Volvo, for example, has achieved a 20% reduction in development times by this means. Another way in which this can be achieved is by parallel development where, for example, work is initiated on construction of the manufacturing plant before the testing of the prototype is completed or the design frozen. The adoption of this *parallel* rather than the traditional *sequential* process increases the financial risk if difficulties are experienced.

The importance of reducing development times has received increasing attention in recent years and a number of major companies are prepared to accept the greater risk in order to gain the potential benefits from an early product launch.

4.9 Summary

In this chapter the relationship between the market place and engineering decisions, particularly in relation to new products, has been explored. The complexities of the many factors involved and the frequent lack of accurate information make it impossible to derive an optimum solution. There are many alternative designs which must be examined, each with different implications for the market. The final choice will depend upon judgements over a number of trade-offs.

Although it is highly desirable that the engineer has a market orientation he cannot be expected to possess an intimate knowledge of the market. For this he must rely upon others who do possess this knowledge – the marketing department. But it is not sufficient merely to integrate the technical with the market, since every decision also has corporate, manufacturing and financial implications. Thus engineering is only one aspect of what must be an integrated decision-making process.

Integration is not easily achieved in practice. Each of the departmental interests will have its own priorities and perceptions. The individuals within the departments will relate more to their own superiors in the company hierarchy than to the needs of other functions. This applies as much to engineers as to others.

Thus integration does not occur naturally; it must be made to happen. Organisational structures and procedures can provide a valuable framework, but they can only facilitate a process which must extend much more deeply. It is the attitudes and actions of individuals working within this framework that create the reality of meaningful integration; only then can the procedures be used to shape the decisions. It will never be known whether the best decisions have been taken, but they will certainly be improved when all involved, particularly engineers, see their own efforts as only one element in what must be a market-oriented profit conscious business.

Technology, engineering design, the market place and the financial environment are undergoing constant change. New information arising in any one of these areas must be communicated swiftly and its implications examined elsewhere in the business. Once again formal systems are a great aid but they are only likely to function effectively with commitment to the process. This must be continuous, not intermittent.

Chapter 5
The engineer and production management

Production is the process whereby materials or components are transformed into finished products. It is the activity where most of the physical work is undertaken. For many engineering companies most of the resources – financial, human and physical assets – are engaged in production. The importance of the effective management of these operations cannot be over-emphasised. This will manifest itself in the cost per unit of the product where small savings can have a significant impact on profit. Engineers, by training and inclination, have a sympathy for the physical aspects of business and natural orientation towards production processes and their management. As a consequence it is not uncommon for engineering companies to become production dominated at the expense of other business activities such as marketing and product development. This chapter is concerned with the role of production within business, and the engineer's role within this function. It must not be forgotten, however, that production is only one part of the integrated activity we call the business; there are many occasions where the optimum business decisions are not consistent with the most desirable production operating conditions.

Engineers are involved in production at two levels: the design of manufacturing equipment and systems, and their operation to produce the end product. The activities range from large one-off projects such as the design and construction of a power station at one extreme to the manufacture of mass consumer products at the other. The differences in the role of the engineering manager in the various types of production system are sufficiently great that they need to be discussed separately.

Another problem in discussing manufacturing management at present is that both manufacturing technology and managerial systems are undergoing a revolution under the impact of a wide range of electronics-based developments. these include computer aided design (CAD), computer aided manufacture (CAM), computer integrated manufacture (CIM), flexible manufacturing systems (FMS), just-in-time, advanced manufacturing systems (AMS) and robotics. Many companies are rapidly moving towards these new systems, which have little in common with the traditional approaches. As a consequence the nature of the production manager's job, the skills he needs and his career pattern are undergoing profound changes. At the same time there are large sections of the engineering industry which are still operating in the traditional fashion, because of either the nature of their operations or their reluctance to embrace technical change.

The word 'revolution' to describe these changes is apt. Success depends upon a complete rethink of the management of the production process. Research has shown that the adoption of an incremental approach to the new technologies is likely to yield only a small proportion of the potential benefits. The reasons for this will be discussed later in the chapter, when it will be seen that many of the traditional problems besetting the production manager are no longer relevant. It will also be seen that the new systems demand a much closer integration between the production function and the business as a whole. These developments underline the importance of two themes which have been stressed throughout this book: the role of the engineer as innovator and change agent, and the need to relate engineering decisions more closely to the objectives of the business as a whole.

Thus we are living at a period when two significantly different approaches to manufacture are coexistent. In what follows we shall commence by discussing the traditional problems of production management, since they are still relevant for much of the engineering industry.

5.1 The role of production

The aim of production management is to produce the right product in the right quantity at the right time at minimum cost. More simply, the objective is to satisfy market demands at minimum cost.

However, in the traditional engineering company it is not possible to achieve all these objectives at the same time, because of the inflexbility of the manufacturing operations. Thus from the production manager's viewpoint the ideal situation is to produce one standard product at a constant rate to enable the optimum use of his facilities, thereby leading to the lowest unit production cost. Rarely is this possible. One is reminded of Henry Ford's statement that customers could have their Model T any colour they liked provided it was black. This production centred approach may be possible where demand outpaces supply, but this condition is enjoyed in few, if any, industries today. Thus it is not possible to meet all the objectives listed above. There must be trade-offs between the various elements.

The major inflexibilities relate to production volume and the product mix. Because of the capital intensity of most production processes costs are highly sensitive to the volume of production. Unfortunately demand is unlikely to be constant; it will vary with the economic and competitive climate and in many industries will also follow a seasonal pattern. The other problem relates to product variety. Because of the high costs of adjusting production from one model to another it is economically desirable to minimise the frequency of such changes, thereby increasing the desirability of large batch sizes.

Faced with these difficulties it is necessary to introduce a degree of decoupling between production volume and the market demand. This is achieved by the holding of stocks of parts or finished goods at various stages in the production process. This eases the managerial problem – but at a cost. It can also be seen that high inventories are attractive to both production and marketing management. For production it reduces the pressures from marketing since their requirements can always be met from stock; similarly it enables marketing to satisfy their customers. This partly explains why inventory levels are often unacceptably high where financial controls are lax. When viewed from the financial perspective of the business as a whole an optimum must be arrived at; the cost of holding inventories must be balanced against additional production costs through more frequent adjustments than might seem desirable when the production costs are considered in isolation, or through the possible loss of a customer owing to the inability to deliver promptly. A number of mathematical relationships for inventory control and the economic size of production batches will be found in any standard book of production management. However, it must be recognised that:

1. This leads to a compromise solution where neither marketing nor production can be fully satisfied.
2. As a consequence friction and pressures can arise between the production and marketing departments.

Production operations can be broadly classified as one-off (or job lot), batch, mass, and continuous (or process). The characteristics of each of these categories and the implications for management will now be discussed.

5.2 One-off production

The simplest form of one-off production can be illustrated by the one-man bespoke tailor. He is unlikely to hold stocks of cloth. His customers will choose their requirements from patterns, from which the tailor can order against a firm order from his customer. His capital investment will be small and he has the ability to perform all the tasks necessary in producing the suit himself. His work-in-progress will be minimal and he may even be able to cover the cost by demanding a deposit from his customer. He holds no finished stock. His prices are high and the customer may have to wait several weeks for delivery. His only competitive advantage is the quality of his product. Nevertheless, there are also constraints within which he has to operate. If his costs or prices are too high he will lose customers. Similarly if the customer wants the suit for a special occasion the delivery time will be important.

The tailor may seem far removed from the large one-off engineering project, say the construction of a power station. In practice there are many differences brought about by the complexity of the latter, but in concept the two have a great deal in common. The closer the engineering contractor can approach the conditions of the tailor the more likely he is to succeed.

First of all it is necessary to obtain the contract. In doing this the tendering company must consider the criteria of importance to the customer, which include:

1. Technical performance
2. Price, including financing arrangements
3. Delivery
4. Confidence in the contractor's ability to meet the contract conditions.

These criteria have been listed in the order of importance which may seem appropriate to the engineer. They may not, however, be seen in the same light by the prospective purchaser. Consider, for example, the case of the supply of equipment to an offshore oil drilling rig, where the company submitting the tender is a successful engineering company and the purchaser is a major oil company. The latter may have great confidence in the technical competence of the engineering company. Price may also be of relatively small importance for a piece of equipment accounting for only a small proportion of a major project. What is of critical concern is that the delivery schedule is met; the customer must have confidence in the supplying company's ability to meet the promised delivery deadline. In order to obtain the contract it is the latter criterion that is vital, and a tender document consisting almost entirely of technical and cost data is unlikely to be addressing the essential issues. The engineer must appreciate that technical performance is only one of the criteria which need to be considered and in some cases may even be of secondary importance. It is necessary to analyse the project from the viewpoint of the client.

Every one-off project is unique. In some cases, such as a power station, it may differ little from previous designs; in others, such as the Thames barrage, it will be completely new. Because of this individuality there is limited opportunity to gain from the experience curve effect of repetitive programmes, and it is therefore essential to get it right the first time. This demands that every aspect must be planned in detail, and possible difficulties anticipated, by the use of planning techniques such as critical path analysis and PERT.

A project will also have a well defined beginning and end. Timing is of the essence and the planning process must ensure that the resources needed are scheduled to meet the needs at every stage. However, because a project is to some extent different from any other there will always be an element of uncertainty. Some degree of flexibility must be built into the planning process to allow modifications to be introduced without causing too much disruption, when unforeseen problems emerge. In general the greater the amount of detailed planning undertaken before the commencement of physical work, the more likely it is that the project will meet its costs and time objectives.

Another characteristic of the engineering project is that the quantity and quality of the resources being used are changing throughout its progress. Unless the company is engaged in a large number of projects so that its resources can be phased to enable a constant workload for all the different types of resource, this presents a major managerial difficulty. This problem is usually overcome by employing specialist subcontractors whose work is phased into the project as and when it is required. This introduces another dimension into the management of a project: the selection, planning and control of resources which are not under the direct managerial control of the main contractor.

The managerial team will be formed for the individual project. It is unlikely that many of its members will have worked together preivously. This means that the project manager must direct much of his attention, particularly at the early stages, into creating an effective team, establishing his leadership and motivating the members of his team. This demands considerable interpersonal skills on the part of the project manager, for which all too often he has had little preparation. Success is likely to depend much more upon his managerial ability than on his technical skills. This task is compounded by the fact that large engineering projects are complex. It is probable that no one completely understands the whole project, and the manager cannot be expected to possess an expertise in the wide range of specialist areas of those reporting to him. Nevertheless, he must weld them together and much of his time will be spent on managing the interfaces between these areas. Established policies and procedures are of much less assistance than in organisations engaged in highly repetitive work.

The main characteristics of the one-off project of concern to the engineering manager that have been discussed can be summarised as follows:

1. The criteria for success must be clearly defined in relation to the needs of the client, both at the tender stage and in the conduct of the project.
2. A high degree of detailed planning and scheduling is essential before the physical work can be commenced.
3. There is always some uncertainty since no previous project will have been identical. Thus flexibility and contingency plans are essential.
4. Attention must be focused on the provision of resources which vary in quantity and type with time. Many of these resources will be provided by subcontractors who are not under direct managerial control.
5. Project management requires the ability to lead, develop and motivate an effective team *ab initio*, and to manage the interfaces between specialist functions.

Management of one-off production

It can be seen that the management of an engineering project consists of two major elements. The first of these is the detailed assessment of the resource needs in terms of materials supply, equipment, and phasing of subcontractors. As we have seen earlier a number of planning and scheduling techniques are available, nowadays computer based. The plan is, however, only as good as the information available. Because of its unique character a number of circumstances will occur during the course of the project causing variations from the original plan. The customer may wish to amend his specification; some design modifications may be found to be necessary; there are almost inevitably going to be some human errors in the planning; and some items may be delivered late. In addition there will often be disruptions caused by weather, accidents or labour disputes.

It is the responsibility of the project manager to ensure that in spite of these many unplanned events the original cost and delivery targets are achieved. Almost invariably the project manager is an engineer. Success, however, will depend upon the exercise of leadership and managerial abilities of a high order. In this context it is worth quoting the Chairman of RTZ, Sir Alistair Frame: 'We can make or lose more money out of managing a project than in many of our operating businesses. . . . Project management requires qualities of leadership rather than managerial skills. . . . Top-class managers with powers of leaderhsip are scarce – there are ten times as many people who could run the project as could build it' (*Financial Times*, 4 September 1986).

The term *project manager* is used to cover widely different responsibilities across industry. In some companies he is regarded as a co-ordinator or progress chaser, ensuring that information is collected and disseminated throughout the organisation. Thus he has no executive responsibility for ensuring that action is taken. In such cases it is necessary to establish who is ultimately responsible for the project; frequently it is only at the chief executive level that all the elements are brought together. This limited interpretation of the role of the project manager usually results in insufficient emphasis on action to drive the project to a successful conclusion.

A true project manager is held responsible for the total conduct of the project. He may be likened to a chief executive for this one aspect of the company's total portfolio of activities. There is, however, one important difference. Although responsible he does not have direct control of the resources employed on the project. He must deal with other functional managers for whom his project is only one of several. The project manager cannot demand priority; he can only negotiate and persuade. When in difficulty he can always appeal to a higher level of authority, but this is not something which he can do too frequently. For the most part he must resolve such problems himself; that is the job he is paid to do. His role is ambiguous and some might claim it to be impossible. Yet there is no escaping the dilemma that a project must have firm leadership if it is to meet its objectives yet must compete with other projects for scarce resources.

If it were possible to plan a project in every detail at the outset, all those involved would know exactly what they had to do at all times. Project management is required because of the inability to plan for all contingencies. Thus the role of the project manager is to oversee all aspects of the project, anticipate problems before they occur and ensure that timely action, including revision of the plans, is taken. It is a daunting task, but essential.

In order to exercise this responsibility a number of questions must be addressed:

1. What types of decision need to be taken in order to exercise project control?
2. How many of these decisions are within the authority of the project manager?
3. What is the project manager's role in relation to decisions over which he has no direct authority? What is his responsibility in relation to such decisions?
4. What information is required to take these decisions in relation to resources, time and cost?
5. What are the appropriate measures for this information?
6. In what form should he expect guidance in assessing time versus cost penalties? Are the relative priorities established explicitly?
7. Does the information enable him to calculate the cost implications of his decisions? If not, what additional information is required?
8. How often is information required day-to-day or in relation to previously established 'milestones'?
9. How accurate does the information need to be? Control information needs to be available quickly and in an easily understandable form; timely information is more important than a high degree of accuracy.
10. What data, the actual figures, are required to provide this information?
11. How are the decisions implemented in operational actions and in modifications to the plan?
12. What managerial system, authority and responsibility, is required?

All too frequently adequate answers to these questions are not provided for the guidance of the project manager. In their absence he must make his own judgements, which may not be in the best overall interests of the business. Take for example a project which is falling behind schedule. The slippage might be recovered by incurring additional expenditure, perhaps by working overtime. On the other hand the project may be allowed to progress at the current rate at the expense of a penalty cost stipulated in the contract for the delay in completion. There is a choice, and what appears to be an operational decision can have business implications. If the criteria are not made explicit, decisions may be taken by the project manager in relation to assumed criteria which are not in the organisation's business interests. Every decision has cost, and consequently profit, implications. But the project manager can only discharge his profit responsibilty if he is aware of the business objectives in relation to the project, and has access to the appropriate information.

5.3 Repetitive manufacture

The term 'repetitive manufacture' has been used to described the three other categories of production – batch, mass and process. In batch production, there are a number of distinct components or products which are made regularly but using all or most of the same equipment. Thus when a production run of one product is completed, manufacture ceases whilst the production equipment is set up to meet the requirements of the next batch of a different product. If the set-up costs were negligible, very small batches would be economic. In traditional manufacture, however, these costs are significant and the size of the batch will be determined by assessing the set-up costs in comparison with the costs of holding inventory.

In mass production a standard product will be made continuously. Many industries combine batch and mass operations and there is a spectrum of different systems. For example, the production line in motor car assembly is normally continuous but can accommodate a variety of model variations, whereas the paint shop can only process batches of one colour at a time.

Process production is characterised by the chemical and oil industries where the plant ideally operates 24 hours daily throughout the year with the unchanging material input and product output.

Although there are detailed differences between these three methods of production, they will not be discussed in detail here. In concept they have a great deal in common, and are in marked contrast to the one-off project in that:

1. The product design is highly refined and production unit costs benefit from the experience curve effect.
2. Major change is infrequent so operating systems and procedures can be established in detail.
3. The resources, men and machines, are constant.
4. Changes in market demand have a major impact on the economics of the production system.

In an engineering project, planning is primarily concerned with deciding what is required, who should provide it, when it is needed and the timing of assembly. The quantities of a particular item are usually small. In repetitive manufacture these tasks are only necessary on the introduction of a new model. After that the main concern of production planning is with quantities and time. How many of an item are required, and when? What number should be scheduled for production? The problem for most industries is that production must be planned and manufacture completed before the orders are received. If the product is not available when the customer requires it he is likely to go elsewhere. This is, of course, a generalisation and can be alleviated by the product design strategy. For example, if the product has a distinct competitive advantage the customer may be prepared to wait, as is the case with a Jaguar or a Porsche but not with a mass market model. In the former cases the company's product strategy helps to eliminate many of the market created production problems associated with the latter.

Production planning is, therefore, concerned with three main features:

The variation of market demand This affects both the total volume and its breakdown between product types. A forecasting system, inevitably largely based upon data from the past, modified by seasonal patterns where appropriate, is essential.

The lead times In many industries the time from the placing of an order for materials and parts to delivery, and the time spent between material delivery and the completion of the product may be several months.

The inflexibility of the manufacturing process Since the capital equipment and the manpower cannot be varied to meet fluctuations in demand, fixed costs are incurred whether or not they are producing. Thus the aim must be to ensure a steady production rate in so far as this is possible.

The planning process itself is usually complex. It must be based upon a market forecast. A number of forecasting techniques are available; of necessity they are highly reliant upon past experience, often giving additional weighting to the more recent past. It may also be necessary to make some adjustment for seasonal demand variations. However, the past may not be an accurate guide to the future, so it may be necessary to incorporate the marketing department's assessment of future demand, often subjective, or any special characteristics which mark a difference between the past and the future. The forecast is for the finished product, which may incorporate several hundred individual parts; these have to be ordered, manufactured and assembled. Each bought-in part or material has its own ordering lead times. Each manufacturing operation will use a given resource of employee skill and equipment and takes a specific time; data on these are normally available. Furthermore, the scheduling of the production for one product has to be co-ordinated with the demands on employees and equipment for parts for other products. The overall aim of the planning operation must be to ensure so far as is possible that the products are available when required and that the production equipment is utilised in such a way that bottlenecks are avoided.

This brief description of some of the most important tasks within production planning illustrates the complexity of what is involved. It would be difficult enough if the information were perfect, and the achievement of the individual elements of the plan could be relied upon. Unfortunately, this is extremely unlikely. Market demands change, some items are delivered late, machines break down, strikes occur and so on. Thus we see that the planning itself is likely to suffer from imperfections, and that even if the plan were perfect it would be unlikely to be fully achieved in practice. A major contribution to the solution of these difficulties comes from work-in-progress or the holding of inventories of completed or partly completed components at various stages in the production process. These serve two functions. Firstly, they enable the planner to

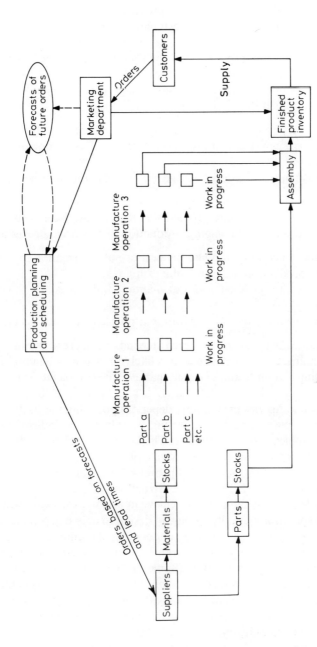

Fig. 5.1 *Repetitive manufacture: planning, scheduling and material flow.*

smooth out the problems of overloading or underloading of facilities. Secondly, they provide a buffer to minimise the effect of temporary disruptions. As with the holding of finished stock these in-process inventories incur a financial penalty. Any idle stock involves an investment which is producing no return. In many engineering companies the time when work is actually being carried out on an item is a very small proportion of the total period between when the raw material enters the factory and when the manufactured part emerges embodied in the finished product. This process is shown diagrammatically in Fig. 5.1. Note that this is a simplified illustration since:

(a) There are likely to be many different types of finished product.
(b) Each product can incorporate several hundred components and sub-assemblies.
(c) Each component may undergo a large number of manufacturing operations.

A vital additional ingredient is the ability of management at the factory level to take action to overcome any problems as they occur. Thus there are three elements in the process of managing production in repetitive operations:

1. Planning and scheduling to make the best use of information regarding the market, the product, and the manufacturing resources
2. In-process inventories to smooth out planning difficulties and to cope with emergencies
3. Factory management to solve operational problems as they occur.

The efficiency which can be achieved depends to a great extent upon the nature of the market and the product. It is likely to be lowest in a multiproduct company serving a volatile and seasonal market. The complexity is much lower in a process industry where the raw materials may be few and the products one or a few. In such industries profitability depends upon a high utilisation rate for a capital intensive plant. Operational management in process industries can only have a limited influence on profitability. The key to success lies with the ability to forecast long-term demand and to design a plant of a size enabling a high rate of capacity utilisation to be achieved. Theoretical calculations of the economies of scale have frequently led to the construction of oversized plants which, because of inadequate demand, never achieve in practice the theoretical economies. Since the plants are usually specific in design to only one product they can be used for no other purpose. Faced by under-capacity utilisation the corporate response is often to attempt to buy market share by reducing prices. However, since the competition is faced with the same problem this normally results in a price war from which no producer is likely to emerge with satisfactory profit margins. An example of this is provided by the oil industry where over-capacity in refining has led to price wars at the petrol pumps, resulting in low profits or losses throughout the industry.

These considerations also apply but to a lesser extent to manufacturing industry, where the equipment and plant is likely to be more flexible. Thus it can, in the longer term, be applied to the manufacture of different products of a similar type. As we have seen earlier the major problems in this type of industry are adapting to short- and medium-range demand fluctuations.

Management of repetitive manufacture
The production manager at the operational level is responsible for ensuring that the planned rate of production is achieved. As with the project manager his role would be largely irrelevant if the planning were perfect and no unforeseen problems occurred. In practice, however, he is under constant pressure in dealing with short-term crises. But whereas the project manager's difficulties are primarily related to the acquisition and employment of varying resources, the manufacturing manager's main concern is to ensure the continued operation of an expensive fixed resource on an hour-to-hour basis. His problems fall broadly into two categories:

1. The rescheduling of production to cope with problems of machine breakdown, labour difficulties and changed demand
2. Ensuring the co-operation of the production employees to maintain uninterrupted operations.

Because of these pressures and the urgency to rectify problems as they occur, the emphasis is upon immediate action. Most of his difficulties relate to matters which may be regarded as relatively trivial in isolation, for example the breakage of a small component on a production machine, but there may be significant financial implications if say the breakage causes a total stoppage of manufacture in a multimillion pound plant. The qualities required of the manager are the ability to take quick decisions and to translate them into effective action. In this environment he has little time, or perhaps inclination, to consider the larger issues of technical change, new equipment, or improvements to the production processes.

Thus it can be seen that there are three main managerial roles in repetitive manufacture, all of which are normally performed by engineers. They are:

1. Planning, programming and scheduling for the current activities
2. Managing the physical activities at the factory level
3. The design planning and implementation of new manufacturing systems to improve current operations or to meet the requirements of new products.

Each of these roles demands different managerial abilities and technical knowledge. In production planning it is necessary to have the ability to comprehend the complexity of the planning task, to understand and use the planning techniques, often computer based, and to pay a great deal of attention to detail. Behavioural interpersonal skills and the ability to make quick decisions, although always desirable, are not key attributes for the successful planner. The shop floor manager must be a man of action. His intervention is usually required at times of emergency when prompt decisions and actions are essential. He, more than any other manager in the company, is likely to be in constant and direct contact with production employees. It is essential, therefore, that he has highly developed interpersonal skills and an understanding of all aspects of employee relations. By contrast the designer and planner of new manufacturing systems requires a wide knowledge of production engineering and the latest advances, not only in individual items of equipment, but also in systems, which are increasingly based upon the many applications of information technology. He may or may not be responsible for the implementation of new systems, which demands abilities akin to those of a project manager.

In a large manufacturing organisation these three roles are almost certain to be kept separate, each with its own staff. Furthermore, in many large engineering organisations change in production equipment or systems occurs only at infrequent intervals, planned to coincide with the changeover to a new product model when the factory will be temporarily closed down. Increasingly, however, it is becoming necessary to introduce change continuously whilst at the same time avoiding major disruptions in production; this can give rise to conflict between the installer of the new system and the factory manager.

Within smaller organisations the planning and operational functions are likely to be clearly delineated but the responsibility for new system design less explicit. As a consequence the imperative of short-term production pressures creates a situation where too little attention may be paid to planning for the future. This is exacerbated by the different managerial skills required by the two roles.

To what extent do engineers working in this system require a business orientation? The majority are working within a well defined hierarchical and functional organisation where their operational role is clear and insulated from the broader market-related and financial concerns of the business. They need to be good managers but not businessmen. An exception might be those involved in the design of new manufacturing systems. A few engineers from a manufacturing background will need to broaden their horizons as they reach the most senior positions. It will be seen later that this restricted role for the engineer needs to be widened with the spread of the new manufacturing systems.

Management of process industry
The managerial situation in process industry is considerably less complex than in manufacture. This can be seen in relation to the three roles described in the previous section.

Planning and scheduling is confined to the provision of the raw materials in the quantities and time required to meet the forecast demand. Once processing commences the various operations are determined by the design of the process plant. Few employees are involved and the operations manager has little or no control over the operations themselves. It has been noted that in other forms of production the role of the project or production manager is to deal with

The engineer and production management 119

planning deficiencies and unforeseen situations; because of the uncertainties of the project and the complexity of volume manufacture the interventions in those forms of production are numerous. This is not so in the process industry, where variations from the standard conditions are rare; when they do occur they may pressage a serious emergency.

In this situation the role of the operational manager is to ensure that the operations are monitored and to take prompt action when necessary. These actions may be infrequent, but the consequences of the wrong decision can be disastrous. The manager's span of discretion is low. He needs little if any knowledge of the business, and his management knowledge requirements are also low; he must manage people, but he has virtually no direct influence on costs.

The third managerial role, that of the design and installation of new plant and operational systems, is normally the responsibility of specialist contractors in process industries.

5.4 Types of production operation: summary

The discussion in the preceding pages has focused on the essential features of the different types of production operation. It has been confined to the concepts and has avoided description of the wide range of managerial techniques that will be found in the production management literature. The main differences between the systems are summarised in Table 5.1.

In the majority of industries the traditional evolution has been from one-off manufacture to continuous or process. The milestones in the development of an industry have frequently been provided by the move from one manufacturing system to the next, e.g. in shipbuilding from individual designs to the constructions of a standard design such as bulk carriers. The strategic skill has been to identify the right time to make the change. The main benefits have arisen from bulk purchasing of items, the experience curve effect in manufacture and the ability to avoid frequent changes in tool setting.

These clear demarcations are now becoming blurred by a number of developments. The potential of the microelectronic technologies, to be discussed later, enables a wide range of products to be designed and manufactured, in the limit with a batch quantity of one, within a continuous production system. In some industries there has also been a trend for manufacturers to move from the batch production of relatively low value-added products (e.g. bulk carriers) to one-off, higher technology, higher value-added products (e.g. offshore oil equipment).

It will have been noted that, apart from project managers of large projects, the role of the engineer in production is largely confined to operational concerns. In the past, until he approached board level there was no need to understand the broader aspects of business. It will be seen in the next section that this is now changing rapidly.

Table 5.1 Main characteristics of production systems

Feature	Characteristic			
	One-off project	Batch	Continuous	Project
Key factor for business success	Quality of project management	Inventory control and scheduling	Production control and production management	Size of plant in relation to long-term demand
Technical uncertainty	High	Medium to low	Low	Negligible
Vulnerability to market change	Low	Medium to high	High	Very high
Resources	Variables	Fixed in short term	Fixed in short term	Fixed
Product variety	Design to contract	Medium	Low	Very low
Managerial span of discretion (production)	High	Medium	Low	Negligible
Specialisation of plant and equipment	Low to medium	Medium to high	High	Very high
Time scale of production management decisions (operational)	Medium	Low to medium	Low to very low	Immediate

5.5 Technological developments and production management

The potential of recent developments in microelectronics has only been partially exploited to date. Furthermore, the technology is still advancing rapidly. As will be seen later the benefits to be derived from the application of advanced technology cannot be fully realised by an incremental adaptation of existing procedures and manufacturing methods. A complete rethink of the production operation is needed in respect of:

1. Organisational structure and organisational development
2. The information system, production planning and control
3. The technology of the physical manufacturing processes.

It should be noted that the ensuing discussion applies mainly to repetitive manufacture. Although there are many significant applications of information technology in one-off projects, their managerial implications are not so fundamental as in repetitive operations.

Let us consider the scenario of a manufacturing company that has exploited the new technologies to the full. A customer orders a product but requires some modifications to the design of the standard item to meet a specialised need. This modification is designed promptly using computer aided design (CAD) and the manufacturing equipment is programmed accordingly. The manufacture is scheduled to take place with a minimum of delay, and the order is delivered the day after it is received. This may appear an unattainable ideal. Yet a very close approach to it is likely to be possible by the late 1980s. As an example of what has already been achieved in applying microelectronic technology one can quote the example of the Japanese machine tool manufacturer Yamazaki. This company invested \$18 million in a flexible manufacturing system (FMS). The results of this investment were a reduction in employees from 215 to 12; in machines from 68 to 18; in floor space from 103 000 to 30 000 ft^2; and in processing time from 35 to 1.5 days. This and other examples indicate that much of the technology required to revolutionise production operations already exists. Notwithstanding this one must not underestimate the managerial challenge necessary to implement the changes; there have been a number of costly failures.

A brief examination of some of the benefits to be derived from the new technologies shows that they can remove almost completely the managerial problems of traditional engineering production management discussed earlier in the chapter, in that:

Production can respond with very little delay to market demands for both quantity and product design characteristics.
The economic batch size reduces to one, since the manufacturing equipment can be reprogrammed almost instantaneously.
Inventories can be abolished owing to the highly sophisticated information processing and planning techniques and the application of just-in-time methods.

Equipment utilisation is improved Already there are some examples of unmanned machine tools operating throughout the night.

Process times are reduced owing to the ability to schedule work more effectively, material will be worked on almost continuously from the time it enters the factory to the time it leaves, rather than spending long periods between operations as work-in-progress or finished inventories.

In total the contribution from new technology in manufacture can be considerable. An MSC report published in 1986 showed that new technology accounted for 60% of productivity growth in US and Japanese industry compared with 15% in changes in the use of labour and 25% from capital investment.

The managerial aspects of these new technologies can be summarised as:

The need for continuous change
Flexibility
Organisational integration
The size and characteristics of the work force
The role of the manufacturing manager.

The need for continuous change

In most engineering companies the basic nature of the production operation has in the past tended to remain unchanged for long periods. New machinery will have been purchased, justified on technical merits and cost–benefit considerations. Even where major re-equipment has been undertaken or new manufacturing systems introduced it has usually been done without a reappraisal of the company's organisational structure. Change, therefore, has taken the form of intermittent disruptions, sometimes associated with temporary chaos, followed by periods of steady production.

Owing to the rapid developments in technology most companies will find that they must adjust to continuous change if they are to keep abreast of the latest advances. To the manager, whose aim is to achieve continous uninterrupted production, all change, however desirable in the long run, is seen as a threat to the achievement of his short-term targets. There are no simple solutions to this dilemma. One alternative is to make the production manager responsible for introducing the change; another is to form a special task force or project team. In the former case the production manager has to reconcile two different styles of management. This is not impossible provided he has received training in the management of technological change. If, as seems likely in the future, there is increased movement between functions, the engineer who is managing production may well have spent a period in R & D which will have given him an appreciation of the technical change process. When a task force approach is adopted the relationship between the project manager and the production manager can lead to conflict.

The technical problems in introducing the new systems are formidable. They can give rise to substantial cost penalties if badly managed. Nevertheless it is only very occasionally that a solution to these technical problems cannot be found. It is the managerial difficulties, often largely overlooked, which are likely to pose a threat to the company's long-term profitability unless they are resolved satisfactorily.

Flexibility

We have seen that traditional manufacturing operations were inflexible owing to constraints in procurement, machine setting, planning and scheduling. Thus the main concerns in productivity improvement related to the *flow of materials*, with management attention directed to such activities as inventory control, work-in-progress, economic batch sizes, elimination of waste etc. The need to optimise production operations introduced inefficiencies elsewhere in the system: product variety was limited, orders were aggregated into batches, and stocks of partly finished or completed components or products were held in various places. This was to ensure that in so far as was possible production operations were decoupled from outside influences.

Just-in-time, computer scheduling techniques and flexible manufacturing systems to a large extent eliminate these concerns. Success now depends upon the effective *management of information*. This enables the organisation to adopt the desirable position of being market driven rather than production constrained. This change of emphasis has important implications for managerial priorities and the organisation structure of the company.

These benefits can only be achieved, however, if the system operates smoothly. In order to do this:

1. The information system design must meet the particular needs of the company.
2. The system must be implemented effectively.
3. When the system is operating, its effectiveness depends upon the accuracy and timeliness of the information inputs. It is highly vulnerable to shortcoming in the provision of appropriate information.

Because of the dominant role of information technology in production management it is clear that the individual manager requires an appreciation of and capability in information technology (IT) in addition to his engineering knowledge. How detailed his skill requirements in IT should be is debatable.

Organisational integration

The needs for continuous change and increased flexibility demand a closer integration of activities throughout the company if it is to be responsive. In particular this applies to top management, production, R & D and marketing. Although it is the information system that is bringing them more closely together, this is of little value if it is not reflected in their decision-making processes.

Where change was infrequent the need for contact between departments was limited and the arm's-length relationships so often typical of interfunctional communications within a functional hierarchical organisation were adequate. A much more organic structure has been found appropriate where the level of innovation is high. Because of the greater uncertainties, creativity, entrepreneurship and organisational structures which bring into close contact those involved have been found necessary to overcome the rigidities of hierarchical systems. This has manifested itself in the adoption of matrix structures and project management. Similar approaches can be appropriate where manufacture is undergoing continuous change.

The need for integration will now be briefly described in relation to corporate strategy and relationships with suppliers, R & D and marketing.

Integration and strategy: Integration must begin at the corporate level. Manufacturing can no longer be regarded as being responsible for the operations necessary for the achievement of a corporate strategy formulated without a significant input from manufacturing. The most senior managers in manufacture, almost invariably engineers, must therefore prepare themselves for a much greater involvement in the company's strategic processes than has been normal in the past.

Recent research into the success and failure of advanced manufacturing technology (AMT) systems indicates clearly that their introduction has frequently been planned solely in relation to technical objectives such as output maximisation or cost reduction. However, in the sanctioning of the investment at the corporate level other criteria, often less easily quantifiable, have been regarded as the major benefits. These might be responsiveness to the market, quality or reduced lead times. This difference illustrates the dichotomy between long- and short-term objectives discussed in Chapter 1. In the present instance, however, this manifests itself as different perceived objectives for the same project at different levels in the organisation. Furthermore, whereas corporate objectives may have played a major part in the original investment decision the research indicates that they are often ignored in the implementation of the system, which has been controlled only against technical objectives. A significant proprotion of the failures in the introduction of new technology into production have been attributed to this cause. The implications are clear. For success this mismatch of objectives must be avoided. This can only be achieved by an explicit manufacturing strategy, formulated with the involvement of production managers, and against which implementation can be measured and controlled. Fig. 5.2 shows these alternative approaches – technical, system and business. Only the last can ensure that the new manufacturing technology is integrated fully to meet the business objectives.

Supplier relationships: The imperatives of a just-in-time system demand a close integration with the operations of suppliers. Recent trends indicate that there is a reduction in the level of multiple sourcing, with reliance being placed upon one supplier who is closely tied into the company's own operations. This has major implications for the purchasing department, which may have to replace an arm's-length competitive tendering relationship with suppliers to one of partnership. This integration can lead to a much more intimate managerial relationship with the supplier, involving both R & D and quality control. This type of relationship is customary with Japanese companies who work closely with their chosen supplier in product design and the establishment of standards; frequently the mutual confidence is such that they dispense with an inward inspection of deliveries.

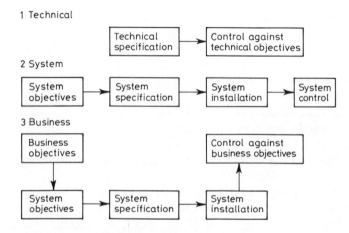

Fig. 5.2 *Implementation of advanced manufacturing technology: planning and control.*
Source: C. A. Voss, 'Implementing manufacturing technology', *Int. J. Op. & Prod. Man.*, 6(4), 1986.

The production/R & D interface: Apart from its role of developing entirely new products and processes, there are two areas where R & D has a more immediate relationship with production. These are the modification of standard products, perhaps in response to a customer's request, and improvements to manufacturing processes.

CAD/CAM enables R and D to design modifications and introduce them into production with very short lead times. This is only possible, however, if the two systems are closely integrated and use a common data base. In relation to production processes the role of R & D should be to aim for continuous improvement through the application of advanced technology and scientific methods in addition to assistance in trouble shooting. This frequently involves the collection of operating data for scientific analysis; often the systems have not been designed to provide these data. Closer integration of the needs of R & D and the initial design of the production processes is essential if the full potential of science and technology is to be exploited.

The integration of R & D and production is founded upon the information system. But this is only a tool, and it will achieve very little unless the people involved at both ends are attuned. Restructuring the organisation and managerial procedures can assist in overcoming interfunctional barriers. Of greater value would be the removal of the barriers. One way to do this is to transfer people, usually from R & D to production. The engineer who has spent some time in R & D will retain his scientific curiosity and personal contacts with R & D staff when transferred to a post in production. Thus what appears as a technical or managerial problem can be alleviated by the adoption of a career progression for engineers involving transfer between functions.

Integration with the market: The ability to respond to market needs has been mentioned frequently in the preceding sections, so the detailed argument will not be repeated. However, it must be stressed that in terms of the success of the business this is the area where most companies are likely to reap the greatest benefits from the new technologies.

The size and characteristics of the workforce
The reduction in the number of employees in manufacturing industry is already significant and has been the subject of much debate. Less attention has been devoted to the skills required by those who remain. At one time it was thought that there would be a high level of deskilling; for example the combination of simple decision-making with manual dexterity would no longer be required. However, the experience of companies at the leading edge of the new technologies suggests that, whilst some of the traditional skills may no longer be relevant, new ones are required, namely:

1. The need for flexibility across traditional skill boundaries.
2. The ability to respond to the unexpected. This demands a workforce with better education and the ability to make decisions and to act in an emergency.
3. The introduction of payment systems linked to the skill and training of the employee and to his ability to take action rather than to the volume of his output, over which he has little or no control.
4. A greater involvement of the worker in decision-making, for example through quality circles.

A number of research programmes in several countries have indicated that employees are more willing to accept technological change than is commonly supposed. Even trade unions, with a few notable exceptions, accept the need for change provided they can extract compensation from management for their co-operation. In many instances it has been shown that the major barriers to change come from middle and senior management.

The changing role of the manufacturing manager
The introduction of the new systems viewed in narrow technological terms

The engineer and production management 127

without modifications to the company's organisational systems lead to a diminution of the manufacturing manager's span of discretion. Many of his traditional tasks are removed and absorbed by the centralised information system. This can reduce his job satisfaction and his career prospects.

The introduction of the new technologies is, however, unlikely to succeed without a major reorientation of management thinking and managerial systems. In this context the demands upon the engineer are likely to be expanded rather than contracted, in relation to both technical expertise and managerial qualities. This implies changes by the organisation and in the training and development of the individual engineer. Amongst the most important of these for the engineer are:

1. A greater understanding of the business implications of technical developments so that he can play a major role in the formulation of corporate and manufacturing strategies
2. An acceptance of a career development involving a greater degree of job rotation between functions
3. The acquisition of a knowledge of the information technologies in addition to his basic discipline
4. An understanding of the elements of technological change and the ability to manage it
5. The adoption of a more participative style of management than has been customary in manufacturing to obtain the co-operation of a more highly educated and multiskilled workforce.

In examining these trends it will be noted that the management of manufacturing is demanding two different personal attributes which are difficult to reconcile. The operational system is becoming increasingly vulnerable to error, particularly in respect of the information fed into it. To overcome this, attention must be focused on the meticulous concern with detail. On the other hand the introduction of change requires creativity and entrepreneurship – characteristics which are often not associated with people who pay great attention to detail. Engineers with both sets of attributes are likely to be the exception. Thus we see two types of role and two types of individual, suggesting that in the future greater attention must be addressed to the matching of the role and the attributes of the individual engineer.

The role of project management is becoming of increasing importance in both one-off and repetitive industries. The project manager is almost invariably an engineer, but the successful discharge of this role, as seen earlier, demands many attributes beyond technical expertise. The training and development of engineers in the wider aspects of management is essential if they are to become successful project managers.

5.6 Quality assurance

The responsibility of the production manager in respect to quality is to ensure that the product meets the specification. In practice, however, there are a number of reasons why there is likely to be a proportion of defective products in any engineering operation, e.g. materials below specification, human error, or wear in machine tools. In traditional engineering companies the monitoring of product quality is vested in a quality control department normally independent of production management.

In some industries every item is inspected. Where safety is involved, such as in aerospace, it is essential that no defective items are delivered. Defective items are scrapped or returned to production for reworking; in either case this involves a cost penalty. The inspection of all items is expensive; thus in most repetitive production companies only a sample of the end product is inspected using statistical quality control methods. In such cases a number of defective items will slip through the net and be delivered to the customer. The rationale behind this approach is that the loss of customer goodwill and the cost of replacing any defective items is less than the additional cost of inspecting every item.

Traditional quality control is based upon the assumption that imperfections in the human and physical production processes inevitably lead to a proportion of defective items. Its purpose is to:

1. Ensure that no or an acceptably low number of defective items are delivered to the customer
2. Identify changes in the defect rate, which can then be investigated; causes can be diagnosed and remedial action taken by production management.

Two developments in recent years have led to a reappraisal of these attitudes towards quality. Firstly, many products consist of a large number of components which are permanently embedded in the end product; electronic products are a particularly good example of this. In such products inspection of the end product could lead to the scrapping of an expensive item through the failure of a minor component. To avoid this it is highly desirable to identify the defective items before they are permanently embodied in the product. Secondly, technological developments, in particular microelectronic control systems, enable a much closer control over manufacturing tolerances. Together these make possible the achievement of very low defect rates, with the ultimate aim of *zero defects*. Associated with this, the term *quality assurance* is increasingly being used. An important element of this is to make the individual operative responsible for the quality of the component on which he is working.

The move from traditional quality control to a zero-defect quality assurance approach is more than just a technical change. Some of the cost implications are self-evident. A manufacturing system which can operate without producing defective items will save the costs incurred through scrap, reworking and lost customer goodwill. This must obviously be a major objective of production management and can have significant financial and marketing implications. These cost reductions will be particularly significant in a mature industry, where profit margins are low.

The implementation of the new system can, however, present considerable difficulties. Let us consider for example the case of a traditional engineering company where employees are paid on a piece rate system. The company's payment policies and the individual's motivation are geared to volume of output, not quality. In most cases the operative knows when he has produced a defective item, but by revealing this knowledge he will incur a financial penalty. Thus a policy which makes him responsible for quality is bound to fail when he is rewarded for quantity. It is beyond the scope of this book to explore this aspect further. The important thing to recognise is that a change which at first sight is a simple technical modification of procedures has personnel, financial and marketing implications. These must be identified and designed into a comprehensive change programme. All too often the engineer acts in isolation, and decisions he takes based upon technical logic give rise to major problems in other aeras. Once again we see that the engineer cannot act in isolation from the total business environment within which he is working.

Chapter 6

Financial implications of engineering decisions

This chapter is not concerned with describing the accounting procedures which can be found in any standard text on the subject. It will, however, examine the implications of engineering decisions in the broader company financial context as well as draw attention to areas where an application of uninformed financial procedures can result in engineering decisions which are not in the best interests of the firm. It will also explore some of the misunderstandings which all too often arise between accountants and engineers.

Let us briefly examine the role of the accountant. Engineers often regard them as obstructive and see them as the operators of the scoreboard, as distinct from the batsmen who score the runs. Or they may be thought of as corporate policemen responsible for discipline in the organisation and restricting the enterprise of those who are working for success. The use of such terms is indicative of the lack of understanding which is prevalent in many organisations. It must be recognised, however, that both accountants and engineers have the same ultimate objectives, namely the long-term survival and profitability of the company. Any internal conflicts prejudice the achievement of this aim.

Until the engineer reaches a senior position he is likely to have limited contact with the finance department as distinct from the accounting department, although in many companies the two activities are normally combined at the highest levels. There is a clear distinction between the two roles, although the edges between them may sometimes appear to overlap. From a financial viewpoint the company can be regarded as an investment. Money is raised from a variety of sources (shareholders, bankers etc.) in two broad categories – equity and debt. The equity holders own the company; they have subscribed money in the past in the expectations of receiving an income from dividends or capital appreciation from a growth in the underlying value of the company. Debt on the other hand is money borrowed in order that the company may use it to earn a greater return than the interest it has to pay on it. The debtors have a prior call on the assets of the company; the shareholders own the residual worth after all debt commitments have been repaid.

The finance director's objective is to raise funds as cheaply as possible through either equity or debt, to maximise the return to the shareholder, and to invest funds in the company to obtain the highest return. He also has to ensure that the cash flows are always sufficient to pay all the bills.

In practice the financial considerations are far more complex. However, it can be seen that there are two aspects with important implications for the engineer.

The shareholders: The company will have a body of shareholders who have invested in it because of their expectations of how it will behave. The stock market value of the shares will reflect these expectations. At one extreme the shareholders may be those who require a stable income, the proverbial widows and children of the stock market. They do not expect the company to make risky investments even if they show promise. At the other extreme are those who have invested for capital growth rather than for income, perhaps high taxpayers. If the company behaves in a way which runs counter to these expectations it may lose its disaffected shareholders, suffer a fall in share price, and perhaps become vulnerable to takeover. Of course, the firm may wish to change and there is no reason why it should not do so provided it is part of an explicit, well publicised and convincing strategic shift. At all times a major objective for the finance director is retention of confidence in the company by the financial institutions. The corporate strategy should reflect these overall financial objectives. It can be seen, however, that this has implications for the major engineering decisions. It imposes constraints. The staid company is viewed with suspicion if it invests in unproven new technologies. Conversely the growth company may lose its stock market rating if it appears to be falling behind in engineering innovation.

Profitability versus cash flow: The engineer normally assesses his projects in relation to profitability, using one of the criteria described later in this chapter. However, although a project may seem outstandingly attractive in these terms, it has no merit if the cash to support it cannot be made available, particularly if there are cost overruns. The failure of Rolls-Royce in 1971 is a classic example of what can happen. This was occasioned by escalation in the cost of developing the RB211 aero engine. The company did not have the cash resources in spite of raising new money from the shareholders through two rights issues. The company was forced into bankruptcy. Yet this has been an outstandingly successful engine, as subsequent history has shown, and although the development costs were appreciably higher than anticipated the subsequent profits have justified it as a project. This is a poor consolation to the shareholders in the original company who lost most of their money.

There should always be more calls for investment than the money available. This statement might be challenged by those in cash rich companies. However, if the purpose of an organisation is manufacture, it might be argued that the retention of large cash balances, unless earmarked for a specific venture, reflects a lack of enterprise. There must be some criteria for deciding which areas of the company and which proposals should be supported. In making this decision priority would normally be given to areas where the anticipated returns are greatest, modified by considerations of risk and corporate strategy.

The mechanism by which the allocations of money are determined is through departmental budgets. These accumulate all the items of expenditure anticipated for the budget period, usually aggregated under the headings of revenue and capital. The total of all the departmental budgets plus any additional items of expenditure must then be compared with the availability of funds. If as is normal there is a shortfall of funds the budgets will be reviewed and modified after a period of further analysis and negotiation.

Thus the budget provides the framework within which the engineer must conduct his operations. The budget is, however, only an estimate. Furthermore, the engineer is not dealing directly with money; his concern is manpower, materials and equipment. The costs of these have to be translated into money terms since this is the only common measure by which they can be aggregated. The accountant is responsible for this translation, comparing the actual with the budget and ensuring that the accounting systems are uniform throughout the company. Although this may appear mechanistic, many judgements are required in practice. These judgements will be based upon accounting conventions and company accounting procedures. On occasions the engineer may dispute the logic of a decision. He may be right but it is essential that common procedures are adopted throughout the company if chaos is to be avoided. There will always be exceptions which do not fit well into these processes, but a degree of illogicality has to be accepted in order to make the overall system workable. This then may be regarded as the scoreboard role of the accountant.

On comparing the actual expenditure with the budget there will almost always be a variance. It could arise because expenditures have not been sufficiently controlled. These expenditures, as we have noted earlier, are not incurred directly in financial terms; they can include overtime worked, changes of materials, machine repairs and many other items. Some will result from managerial failures, others from unforeseen circumstances. It is essential, however, that the variances are contained if overall financial control is to be maintained. But how is this to be achieved? In some companies the term 'controller' is used for the chief accountant, implying the policeman role by which a discipline must be imposed on those who are responsible for incurring the costs. The engineer would respond that this is unrealistic since the accountant is only thinking in financial terms and lacks an understanding of the complexities of the real world of men, materials and machines.

Thus we see the seeds of conflict. The situation as described above paints the picture of the stereotype accountant. Fortunately most industrial accountants today have a greater understanding of the realities of operations. Nevertheless, differences do occur which are real and can only be resolved by mutual understanding and education. The enlightened accountant – and there are many of them – takes upon himself the role of aiding the engineer from the time the

Financial implications of engineering decisions   133

budget is put together to all stages of engineering management which have an accounting impact. Nevertheless it must be recognised that the prime role of the accountant must be collecting and using financial data, whereas the engineer is concerned with selecting the appropriate figures to make the best decisions.

This mutual trust is of particular importance during the compilation of the budget. The engineer, in his drive to avoid a negative variance, is inclined to build in contingency allowances so that when one item of expenditure rises it can be covered within his overall budget. On the other hand the accountant desires to minimize the total size of the budget and will seek justification for the detailed figures in the budget. Since the figures are estimates they must be based upon judgements. Thus the process can become an elaborate exercise in game play and negotiation. The aim, of course, should be to obtain a figure which is reasonable in relation to the nature of the operations and the uncertainties. In this climate an outcome close to the budget may reflect the ability of the engineer to negotiate a liberal figure rather than his capacity to control a tight operation. There are no simple procedures whereby these differences can be reconciled. An adversarial approach is not the answer. Accountants and engineers should be on the same side, but engineers will not be convinced of this if they believe that the budget will be used as a main criterion for assessing their performance as managers. These problems can only be overcome when engineers and accountants can work together with an understanding of each other's objectives and problems. This does not come automatically.

Summarising this introduction, it is clear that:

1 Engineering decisions are constrained by corporate financing policies.
2 Potentially profitable projects can be limited by cash flow considerations.
3 The budget is the main mechanism for aligning engineering activities with the company's financial management.
4 The budget provides the tool for financial control through comparing actual outcomes with the estimates.
5 The engineer requires some knowledge of accountancy principles, and an understanding of the accountant's corporate role.
6 The accountant needs an understanding of what the engineer does and a sympathy for his problems.
7 A good relationship between the accountant and the engineer is essential, and both should seek ways of establishing it.

6.1 Capital budgeting

The acquisition of funds for investment in engineering projects is an area where the engineer becomes most closely involved in financial matters. Money is required for a specific project, and it must be shown that this is a worthwhile use of the company's resources.

Projects can be categorised under three broad headings:

Research and development In most cases the project will be undertaken within a fixed annual budget. The problem is to choose the constituents of an R & D portfolio. Only rarely will a large project be funded separately.

Major capital projects There are two types of capital projects: those undertaken for the company itself, and those undertaken on a contractual basis. The first might involve the building of a new manufacturing or processing plant which will be used in the production of the company's own products. The second type is where the main activity of the company is contracting – the building of a power station or a steel plant. Although the contract price will be a matter for negotiation by the finance department, taking into account a range of non-technical considerations, an estimate of the costs is necessary to provide a basis for tendering.

Equipment replacement These items are usually smaller investments in plant and machinery to improve productivity or quality.

The major problem in assessing projects is estimating what the costs and benefits are likely to be. In the majority of cases the projects may bear little comparison with anything which has been done previously. The uncertainties are greatest in R & D where it is difficult to assess not only the costs, owing to the inherent uncertainties associated with new technology, but also the benefits, which depend upon the size of the market for the new product. For major capital projects the uncertainties relate more to the cost since it should be possible to compute the benefits within a reasonable margin. In the more routine investments it should be possible to estimate both cost and benefits with greater accuracy since there are likely to be fewer unknowns.

Estimating accuracy

The value of any financial analysis of a project must depend upon the accuracy with which the costs and the benefits can be estimated. Unfortunately the uncertainties in most projects are so great that there is often a serious discrepancy between the estimates and the final outcome. Where this relates to a major project it can lead to financial disaster, including bankruptcy of the company.

In order to assess the magnitude of this problem it is useful to examine a number of examples of what can happen in practice. The difficulties in R & D are well chronicled. A number of research studies have revealed that both the costs and the duration of projects suffer from major escalations. Cost outcomes of two or three times the original estimates are not uncommon. In some well publicised cases, for example the Concorde, the outcomes may be as much as ten times the estimates. If this seems a major problem it is often overshadowed by the inaccuracy in the estimate for the market for the product, which in some cases may prove to be non-existent.

Major capital proejcts suffer from the same difficulty, as evidenced by the Thames barrage and the nuclear power station construction programme. This is an international problem which cannot be attributed to the local difficulties, for example trade union militancy, in any one country. The Alaska pipeline, due for completion in 1972 at a cost of $900 million, eventually cost $8000 million by the time it was commissioned in 1979. In Japan the Seikan tunnel connecting the islands of Honshu and Hokkaido was estimated in 1954 to cost 60 billion yen for completion in 1964; construction did not start until 1971, by which time the cost estimate had risen to 200 billion yen; by the time of its completion in 1988 the cost is expected to have risen to 680 billion yen.

These problems also occur, though to a much lesser extent, in more routine evaluations of manufacturing equipment. In such cases the cost of the equipment itself and its theoretical performance are rarely the explanation. It is the costs of installation and commissioning which are underestimated. Furthermore, the benefits are assessed in relation to a rate of utilisation which it may not be possible to achieve in practice, not through shortcomings in the equipment itself but through limitations in total production systems.

Any financial analysis which ignores these realities leads to decisions which at a later date may be bitterly regretted. The dilemma is that:

1 There is no avoiding the necessity for projects if the company is at least to maintain its current position, or to plan for future growth.
2 Because of inherent uncertainties, estimates for both costs and benefits cannot be expected to be accurate and the errors almost always have an optimistic bias.
3 The errors will be the consequence partly of managerial failures, an inadequate analysis or poor managerial control, and partly of genuine uncertainties.

The question is what can be done to alleviate the consequences of these inaccuracies. The fact that they cannot be entirely avoided does not mean that projects must be regarded as a gamble over which management has little control. The responses of management should include the following:

Realism The inherent uncertainties should not be ignored and hidden by the sophistication of the evaluation procedures. Wishful thinking that it will not happen this time can be the prelude to disaster.

Contingency allowances The financial direction of the company must have contingency plans to raise additional funds in an emergency. These can be activated if they are needed. The time to raise the additional funds must be when the first signs of difficulty appear and the company is still in a strong bargaining position. Delay until the crisis has struck puts the firm in a weak negotiating position.

Portfolios The nature of uncertainty is such that it is not possible to know in advance which project will run into trouble. If it was possible, the problem would not exist. However, in a portfolio of projects some will perform better than others. Statistically, the actual/estimate ratio for a portfolio will be less uncertain than for any one project. This approach is applicable to most R & D departments, but is of no value to a company engaged in only one or two major projects.

Post audit Estimating errors can be reduced by learning from the experience of past projects. The sources of difficulty in earlier projects can be analysed. This can help in two ways. It can identify and attempt to rectify managerial weaknesses. It can help to form a judgement on the magnitude of some of the most common errors. This is only possible, however, if the data are available. As a generalisation, few companies or managers attempt to learn from their own history.

Rigorous analysis This ensures, as far as is possible, that all anticipated costs are taken into account and nothing of importance is overlooked.

Detailed planning The full extent of a project cannot be assessed until it has been planned in detail. This will reveal activities which could not have been evident at an earlier stage. The greater success of many Japanese projects is often attributed to the meticulous planning which precedes the initiation of physical work. Premature action can lead to delays and costs resulting from learning as the project advances; in some cases it might involve the destruction of work already done. The classical example is the removal of part of a new factory wall because the equipment won't pass through the door.

Project control The plan identifies what should be done, its costs and its time scale. An information system is required to provide management with a measure of how much has been achieved. The difference between the plan and actual progress gives an up-to-date indication of any departure from plan. This enables action to be taken at two levels – the company and the project. At the corporate level, particularly for a large project, it gives timely warning of the need for additional funds. It may on occasions lead to a re-evaluation and a decision to cancel the project. It must be accepted that as a project advances the quality of the information improves and might indicate that it is no longer financially viable. Projects are cancelled for these reasons. If this is to happen the decision should be made as soon as possible. Frequently this is delayed, leading to unnecessary and avoidable expenditure. At the project level the control data provide the manager with the information he requires for action. The actions take a variety of forms: it may be necessary to make a major review of the plan, or to change the sequence in operations, or to increase overtime. It will normally be unlikely that any managerial action can ensure that the original project objectives are met. The choice facing the manager is likely to be whether to spend additional money to regain the original completion date or to amend the date at a lower cost penalty. In order to make this choice the project manager needs to have a clear indication of company priorities. These may be set in relation to market considerations, for example how critical late delivery will be for future custom, or financial aspects such as penalty payments. The choice should not be (although it often is) left to the discretion of the project manager, who cannot be expected to be fully aware of all the corporate implications.

6.2 Financial evaluation techniques

In spite of the difficulties in estimation, no project will be authorised without a financial justification. For the more tangible investments where a reasonably accurate calculation can be performed this may be the sole criterion for the approval. However, for most projects there will be other criteria, some difficult to quantify, which will influence the decision. The most commonly used financial evaluation methods will now be described briefly. They will be presented in a simplified form; for a deeper discussion reference should be made to texts on management accounting.

Payback
This is the period for the cost savings or profit benefits from the project to pay for the investment, i.e.

$$\frac{\text{total investment cost}}{\text{annual benefit}} = \text{payback in years}$$

This is a simple calculation which is normally only used for relatively small projects with a short payback period, say two years or so. In these circumstances the investment might be justifiable in times of economic uncertainty when there is a reluctance to provide funds for long-term projects, however attractive they might appear on paper, or when the investment itself appears vulnerable to future development, for example technological change. For example, it may be that the company is planning to close a factory within four years. In normal circumstances any investment in it would not be countenanced. Notwithstanding this, a two-year payback project to improve performance in the intervening years could make economic sense.

Benefit/cost
In this calculation the total benefit is divided by the total cost to provide a simple yardstick of the value of the project. It takes no account of the time taken for the benefit to be achieved. For this reason it is usually inappropriate for making the final decision to approve a project. However, it does provide an easily used method for comparing alternative project proposals if the time scales are comparable.

Return on investment (ROI)
This is the annual return on the investment by which it can be compared with

alternative uses for the funds. There would, for example, be little sense in investing in a project with an 8% ROI if the company is borrowing money at 10%. The choice of the appropriate rate to use is a matter for much discussion in the financial textbooks. It might be argued that any rate of return greater than the interest rate that the company has to pay on new borrowing yields a net benefit to the company, although it ignores the risk inherent in any industrial activity. Alternatively, it might be considered that the rate of return on existing activities should be used as the basis. This could be criticised on the grounds that the company should be aiming to increase its profitability, and that the existing activities include some which would not be countenanced if currently proposed. There are also a number of other alternatives which need not be discussed here. Most companies, however, establish a target or 'hurdle' rate which must be achieved by any new proposals. The rate itself will be established by the finance director, and the engineer need not be concerned with how it is set. Nevertheless it can be seen that the financial criteria that a project must satisfy are closely linked to corporate policy.

Because the economic environment of the company is changing, so will the hurdle rate. This is the mechanism by which government intervention in interests rates stimulates or discourages investment. The engineer must be aware of these changes since it could mean that a project which has been rejected on one occasion may become acceptable at a later date.

The hurdle rate is, as the name implies, a barrier that the project proposal must clear. This does not mean that it will necessarily be approved; it merely establishes its eligibility to compete with other proposals for the limited funds available.

Discounted cash flow (DCF)

Money has a time dimension. £1 now is preferable to £1 in a year's time. But what sum of money offered in a year's time would be equally acceptable to £1 in the pocket today? The rate, called the *discount rate*, connects the two. Thus a discount rate of 10% means that £1 today is equivalent in the judgement of the company to £$(1\cdot1)$ next year, £$(1\cdot1)^2$ the year after, and so on. The calculation is like compound interest in reverse. One is concerned with the present value of money in the future; consequently the present value of £1 next year is £$(1/1\cdot1)$ = £0·909, the year after £$(1/1\cdot1)^2$ = £0·826, the following year £$(1/1\cdot1)^3$ = £0·751, and so on.

The effect of discounting at 10% can be seen by comparing a project on the basis of benefit/cost and DCF. The estimated costs are £10 in each of the first two years, and estimated benefits are £20 in each of years 2 and 3.

$$\frac{\text{benefit}}{\text{cost}} = \frac{20 + 20}{10 + 10} = 2$$

$$\text{DCF} = \frac{20(0\cdot826 + 0\cdot721)}{10(1 + 0\cdot909)} = 1\cdot62$$

Financial implications of engineering decisions 139

In calculating the DCF all the cash flows both out and in are converted to the present value using the appropriate discount rate. It is seen from the example that the effect of discounting is to reduce the desirability of the projects. This will always happen since the investments precede the returns. When used in this way, applying a given discount rate, one arrives at the *net present value* (NPV). Alternatively, one can find the discount rate which would equalize the outflows and inflows, the *discounted rate of return*. Using NPV any project with a value of greater than one, using the corporate discount rate, would be acceptable. Alternatively by finding the discounted rate of return a comparison can be made with the required corporate rate.

The time dimension of money is a factor which cannot be ignored. It is normally used in assessing all large capital projects. It also provides a valid basis for comparing projects with similar time frames. On the other hand it cannot be used as the sole criterion where projects with widely different time frames are compared. Let us examine the effect upon a portfolio of R & D projects. Because discounting favours projects with a shorter payback period, these have an appreciable advantage when comparing short- and long-term projects. Short-term projects, however, are likely to be small; they are also more likely to be process than product developments since the former usually recover their investment most quickly. Thus the application of DCF in the selection of the portfolio could distort it in a direction contrary to that indicated by the R & D strategy. Furthermore, if the technique is applied blindly these undesired distortions may be introduced without sufficient recognition. It must be stressed that, important as the financial analysis undoubtedly is, it should be regarded as only one of the inputs to decision-making. The manager cannot rely upon any technique to yield a simple solution to his problems. Life is too complex.

Risk and uncertainty
None of the evaluation techniques as described above explicitly takes account of uncertainty with its associated risk. A recognition of the uncertainty associated with all new projects can be allowed for in the aggregate by adjusting the hurdle rate upwards. By setting a higher target it is hoped that the outcome will yield an actual return which is better than, and certainly not inferior to, what is currently being achieved by existing investments.

This does not, however, enable the uncertainties surrounding individual projects to be compared. The uncertainty profile of any project may be expected to follow the normal distribution. In Fig. 6.1 the profiles for two projects are compared. It can be seen that they exhibit very different characteristics. If the criterion for selection were solely the expected ROI then project A is clearly preferable since the median of the curve (Y) is higher than that for project B (X). There is also a probability of a very much higher outcome. On the other hand the outcome could be much less, and could even be a loss (negative ROI). It is a risky project with the prospect of high returns. By contrast project B is much less risky; there is no likelihood of it incurring a loss, but neither is it likely to achieve a high return. Which of these projects is preferable? Who should make the choice?

There are no easy answers to these questions. They should not, however, be left to the engineer to decide; he needs guidelines. In general the choice will be affected by the corporate risk propensity. Some companies are more willing to take risks than others. In some organisations a high risk might be acceptable provided there is a low probability of a loss. Furthermore, the attitude to risk may vary according to the economic circumstances of the time. The problem is accentuated by the difficulty of expressing the corporate attitude in precise terms. As a consequence many companies do not address the problem with sufficient clarity for those who have to take the decisions.

Often there is a tendency to ignore the uncertainties. When presented with a risk profile the manager will want a simple solution. He will be inclined to ask: what is your best guess? The best guess is the expected ROI, which ignores the inherent uncertainties; thus information which is a valuable input to the decision-making process is put aside.

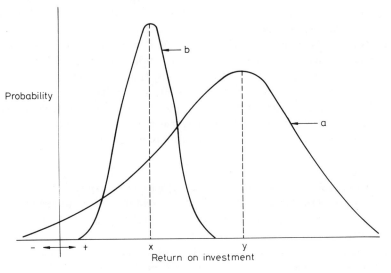

Fig. 6.1 *Project risk profiles.*

The curves shown in Fig. 6.1 can only be derived subjectively. They merely represent the judgements of those who understand the project best – the engineer in respect of the costs and the marketing man for the benefits where the project is for a new product. Like so many techniques the value of the curves is that they force the explicit consideration of risk and enable the individual judgement to be represented in a visual form. When this is done they provide a basis for discussion, possible modification, and a consensus of views.

It can be concluded that:

1 Uncertainty must be assessed, particularly since it is often high in engineering projects.
2 Guidance on the corporate attitude to risk should be given.
3 Risk profiles for individual projects should be drawn based upon the judgements of those best able to provide them.
4 Corporate risk can be reduced by constructing a portfolio of projects with different individual profiles.

Review
There must be a logical framework to provide a basis for investment in engineering projects. Since their ultimate justification must be the contribution they make to the economic welfare of the company, a financial evaluation is essential. Although a rigorous financial analysis must be undertaken it is worthless if it

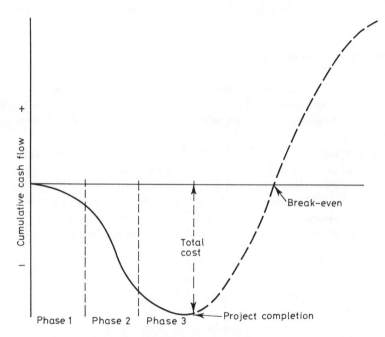

Fig. 6.2 *A typical project cumulative cash flow curve.*

does not reflect all the realities of the situation. This is where the difficulties commence, since a review of engineering projects of all types indicates that initial estimates are often wildly inaccurate. To the extent that they can be improved, managerial attention should be focused. Desirable though this may be it can only partially alleviate the problem. The nature of an engineering project is such that there will always be residual ignorance. If the job has not been done before there will always be some aspects which cannot be anticipated, however great the detailed planning.

In summary it can be concluded that:

1 Wishful thinking about the accuracy of estimates can lead to serious trouble; there must be *realism*.
2 A financial evaluation must be carried out, taking into account expected costs and benefits; the time dimension of cash flows; the availability of funds; and the risk and uncertainty.
3 The financial evaluation is only an aid to the decision, which must also reflect other considerations, often difficult to quantify. In this the strategic objectives are particularly important.
4 The criteria for financial acceptance can vary with time, and are influenced by general economic and corporate financial circumstances.
5 Improved information as the project progresses may necessitate a re-evaluation of its viability. To enable this an efficient information system is essential.

6.3 Cash flow and project design

A project is completed by the employment of various resources – people, materials and equipment – during its development or construction. These represent expenditures which vary with time. The cumulative cash flow for a typical project is shown in Fig. 6.2.

In the previous section it has been seen that it is difficult to estimate the magnitude of the cash flows and the total in advance. Although this makes it difficult to calculate the size of the curve, the shape of it is not likely to depart significantly from that shown. Within certain limits the curve can be modified, since it is a representation of all the managerial decisions taken during the course of the project. With this in mind it is possible to examine the curve and ask how it might be modified to improve the financial outcome, and what the implications of this are for engineering management. It must be remembered that the overall aim is to maximise the ROI from the project.

Maximising the ROI is not the same thing as minimising the cost. This would only be so where the return in unaffected by a cost reduction, which is rarely the case in practice unless the cost can be reduced by the removal of inefficiencies (which should not have been there anyway). There are two main means for improving the return which are within the control of engineering:

1 To increase market share by technological improvements to enhance the performance or reduce the product price
2 To extend the market life through bringing forward the launch date or delaying obsolescence through technical excellence.

Although the above has been written in relation to a product development, similar considerations apply to other forms of engineering project. It can be seen that neither of these objectives is likely to be attained without additional expenditure, i.e. increasing costs. These costs are justifiable if the rate of return from them is equal or greater than the original ROI. This is frequently the case, and it is unusual for the minimum cost to be associated with the optimum ROI. This does, of course, raise problems since the organisational pressures are usually focused on cost reduction. This can be extremely short-sighted. A project is of no value until it is completed and earning a return. It must be accepted that a crash programme is often expensive and is generally unwise. This is not the same as spending limited additional funds to accelerate the project within reasonable limits. A tight programme often brings the additional advantages of strict managerial control and a more highly motivated project team.

The curve can be seen to have three phases. At the beginning – the planning stage – the costs are relatively low. They then rise rapidly as materials and equipment become involved (i.e. construction project), with pilot plant or prototype testing (i.e. chemicals or aerospace) or during installation (i.e. manufacturing equipment). The rate at which the expenditure rises in this phase is often grossly underestimated. One of the reasons why so many entrepreneurial high-technology companies fail in spite of a potentially successful product is that the cost of moving from feasibility to marketable product has not been appreciated. It is a useful exercise for engineers to examine the data from earlier completed projects to obtain the dimensions of a typical cash flow curve. The results are likely to come as a surprise. During the final phase the curve flattens again. The direct expenditure is relatively low but the time can be a significant portion of the total. The real costs are, however, appreciably higher since the cumulative investment made to date is earning no return.

It is during the first and last of these phases that time can be bought most cheaply. At the beginning of the project 'there is all the time in the world' and there may be no sense of urgency. This must be avoided; every effort must be made to explore ways in which the additional application of resources can bring the preliminary stages to a hasty conclusion. This does not imply a neglect of detailed planning. Deficiencies in planning cause delays in the next critical and expensive phase. In general many organisations leave too much to be sorted out later through insufficiently detailed planning. The technically important activities are mostly completed in the second phase. At about this time a new major project might emerge to which the project manager may be transferred, often resulting in a loss of informed and committed management effort. Delays accumulate and the project drags on whilst the final stages are completed, e.g. a nuclear power station standing idle whilst it is being decorated, or a ship waiting for its furnishings. Plant commissioning often takes much longer than it should, often through inadequate planning.

In conclusion it can be seen that:

1 A project should be planned and managed in relation to the cash flow curve to maximise its discounted rate of return.

144 Financial implications of engineering decisions

2 Engineers should be aware of the shape of the cash flow curve typical for their activity.
3 Cost minimisation is not the prime consideration.
4 Urgency is essential from the point of project initiation.
5 The tempo must be maintained until the final completion of the project.

6.4 Finance and engineering decisions

The main purpose for which the engineer requires accounting data is to assist him in making decisions. Because of the complexity of accounting conventions designed for other purposes it is essential that the engineer considers carefully what figures should be used in arriving at his decision. A lack of understanding and the use of inappropriate information frequently lead to poor decisions. These can be avoided if he bears in mind that:

cost of decision
= total future cost − future cost borne in absence of decision

A number of decision areas will now be discussed where poor decisions can be made when this simple relationship is lost sight of.

Sunk costs
Money spent in the past is irrelevant to decisions for the future. Good money should not be thrown after bad. It does not matter whether £10 or £6 million have been spent on a project to date when deciding whether it should continue or be abandoned. That money has gone irretrievably unless some of it can be salvaged or put to other uses. This is a simple truism which is often forgotten. The reasons for this are many: an emotional attachment to the project; a reluctance to admit that a mistake has been made; or the erroneous belief that the money is only wasted when the project is abandoned.

It takes courage to abandon a project once it is commenced, but if its future prospects are unsatisfactory then there is no avoiding the harsh decision. The considerations involved can be seen from the examples in Table 6.1. All three projects in the table are comparable at the outset. However, when new estimates are made at time t they present considerable differences. Project A has more or less progressed to plan, although there is a slight increase in the estimated cost and a reduction in the benefit. Project B is overrunning in cost by a large amount; the eventual benefit has also been substantially decreased. At the end of the day it will only recover about 76% of the investment made in it. Obviously it was a mistake to initiate it, but should it be cancelled? Comparing the new expected benefit with the new cost estimate shows that the return on this additional investment is the same as the original figure. Remembering that the cost of 23 units already incurred is a sunk cost, it will be preferable to proceed with the project. Project C is different. It can be seen that there will be no gain

Financial implications of engineering decisions 145

Table 6.1 *Estimate revision in relation to sunk cost*

Project	Original estimates			Estimates made at time t				
	Cost 1	Benefit 2	Benefit/ cost 3 = 2/1	Cost to date 4	Future cost 5	Total cost 6 = 4 + 5	New estimate of benefit 7	Benefit/ future cost 8 = 7/5	New estimate for benefit/cost 9 = 7/6
A	20	40	2	15	6	21	38	6·33	1·81
B	20	40	2	23	14	37	28	2·00	0·76
C	20	40	2	15	20	35	20	1·00	0·57

from the additional expenditure of 20 since at best the benefit will only cover the cost. It must be cancelled.

Depreciation

The full cost of a piece of equipment is incurred when it is paid for. However, it has a continuing value and might have a working life of ten years or so. If the total cost were charged to the profit and loss account in the year of purchase it would introduce a distortion into the annual profit. The aim of the accountant should be to present figures which enable one year's performance to be compared fairly with another's. This would obviously not be the case if the profits were reduced considerably in one year, owing to the purchase of an expensive item of equipment, and little the following year. Thus the accountant will spread the purchase price over the estimated life of the equipment. His calculations may also have to take into account any tax concessions for investment which may be in force at the time. It must be recognised that depreciation is an allowance in the books of the company and does not reflect the real worth of the equipment. In theory it might be revalued every year, but this would be an expensive exercise and not very useful.

There is a great deal in common between sunk costs and depreciation. The difference lies in that most items of costs included in sunk cost calculations relate to revenue items such as labour. Depreciation is applied to capital items of a tangible physical nature. Because of this they are handled differently by the accountant. These differences should not be allowed to influence engineering decisions. If, for example, a year after the purchase of a machine a new machine with a much enhanced performance appears, it should be appraised on its own merits using the scrap or resale value of the existing machine rather than its depreciated value.

At one time many companies capitalised their R & D expenditure. The logic behind this was that it is an expenditure for the future which bears little direct relationship to the annual operating accounts. Thus it seemed sensible to spread it over a number of years rather than to penalise current profits. This could, of course, be justified only if it could be guaranteed that it would terminate in a profitable product or process. Because of the higher risks of R & D, accounting conventions normally require it to be charged in the year in which the expenditure is incurred.

In all these accounting decisions fine judgements have to be made. For it might be argued that there is little difference between R & D and investment in a high-risk capital project, say an ethylene plant which can only be used for one purpose and only used profitably provided there is a market for ethylene. It is for these reasons that the accounting conventions are required to impose some standardisation between companies' practices and to reduce the area of judgement as far as possible.

It is often difficult for the engineer to understand how the accountant uses his figures. The important thing to remember is that they are produced for purposes different from the decisions faced by the engineer. They should not be allowed to introduce distortions in these decisions which will lead to a conclusion which is not in the best interest of the company.

Thus in considering a new investment, say in machinery, the size of the new investment is the cost of the machine less the value of what it is replacing. But the value of the old machine which is relevant is not the depreciated book worth but its actual scrap or resale value. When these are different, as they usually are, the accountant must make a book adjustment in relation to the disposal price. But in the evaluation the engineer must ignore the current book value and estimate what it is actually worth, rather than what some years previously the accountant assumed it would be worth now.

Fixed versus variable costs
It is convenient to regard costs as falling into two categories – fixed and variable. Fixed costs are those which do not vary with the production rate or the individual project. Some of them will arise outside the engineering department – administration, catering and transport. Others will arise within the engineering department – supervision and heating. In many companies the majority of fixed costs are not directly attributable to a particular project or product; there are, however, exceptions such as specialised machines, tools or laboratory equipment which cannot be used other than for one purpose. Variable costs, on the other hand, are those which vary with and are directly attributable to the rate of production or the engineering project.

The division of costs between these two classes is a useful concept. It provides the basis for drawing the break-even chart (Fig. 6.3). It can be seen that break-even is analogous to payback except that it is expressed in terms of production volume or sales rather than time. Note that the ratio of fixed to variable costs is to some extent dependent upon product design decisions, which must reflect an assessment of expected sales volume and the associated manufacturing costs.

On closer examination, however, the real situation is not so simple as might be thought. The distinction between the two classes of cost lies in the time scale of the resource being considered; in the long term all costs are variable. Labour costs, for example, are normally regarded as variable. If a project demands the recruitment of new staff then those costs are directly associated with it and can be categorised as variable. Labour cannot these days be hired and fired to meet the short time needs. There are recruitment costs, redundancy payments and other legal constraints. Thus in the short to medium terms they have more in common with fixed than variable costs. For example, a research engineer is likely to have a reasonably secure future irrespective of the project he is engaged on; is he a fixed or a variable cost? If there is no other project available for him and there is no question of his being made redundant, then his employment costs must be borne by the company whether or not that project is supported. It can be seen that whether a cost is truly fixed or variable depends upon the circumstances of the time. The accountant must be consistent. On the other hand the engineer, when considering the desirability of a particular activity, has to consider (as has been stressed earlier) the additional cost which must be carried by the company as a consequence of his decision. If buildings, equipment and labour are available, there is no intention to dispose of them, and there are no alternative uses, then these do not represent an additional cost to the company as a result of his decision. Such considerations do not lend themselves to an easy resolution, but they cannot be ignored.

Another problem arises from the allocation of overhead costs. Often in a standard cost system these fixed costs are allocated by the accounting department. The argument is that these costs have to be covered by the wealth creating activities each of which should bear a proportion. Often these are allocated as a percentage added to the direct labour cost. Let us examine how this may affect a purchase decision:

	New machine (£)	Old machine (£)
Annual depreciation	3000	900
Maintenance	600	300
Direct labour	2000	4000
Overheads (as 100% of labour)	2000	4000
	7600	9200
Annual production (no.)	600	600
Cost/unit	£12·67	£15·33

These calculations indicate a significant saving in production costs by purchasing the new machine. However, if the machine does not result in any savings in the overhead costs their total cost to the company is reflected by the decisions, and the apparent saving of £2000 is illusory. Removing them from the calculation yields the following: cost of new machine (without overheads) £5600; cost of old machine (without overheads) £5200. Therefore the decision to purchase would involve the company in an additional annual cost of £400 rather than a saving of £1600. In company terms the investment could not be justified.

This problem is recognised by many accountants nowadays, who hold the view that the engineer should only be accountable for those costs over which he has direct control. Thus his activities can be measured in relation to the contribution they make to the company's overheads.

Capital intensity
The engineer always has a choice between alternatives when considering a project. Some approaches are more capital intensive or require a larger investment than others. At one extreme may be an unsophisticated design suitable for manufacture on standard machines, perhaps involving a degree of hand fitting. Alternatively additional design effort may be applied to the development of a

product requiring a special manufacturing process. In the first case the fixed investment in design and machinery is low but unit production costs are high. In the second these proportions are reversed.

The implications of this choice can be seen by reference to Fig. 6.3, from which it is apparent that the low-fixed-cost alternative A has a lower break-even point, but that B yields a higher profit margin if the expected sales volume is high. It is obvious that project A is the lower-risk alternative but sacrifices profit if the outcome results in high sales volume.

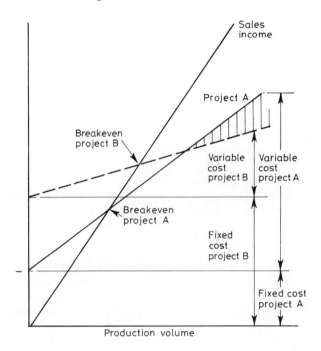

Fig. 6.3 *Break-even chart and the effect of alternative design configurations.*

Although the engineer is deeply involved in these decisions it is clear that they should not be his alone, since they require an agreed market forecast, affect the size of the technical investment, and can have a significant impact on company profits.

The experience curve
As the cumulative production volume for a product rises, so does the unit cost fall. This is a reflection of both technical and managerial learning. When plotted on a log-linear scale a wide range of different products exhibit a straight line relationship with a slope of between 70% and 80%. This is termed the experience or learning curve (see also Chapter 4). For a 70% curve this means that the second item produced costs 70% of the first, the fourth 70% of the second, and so on.

Fig. 6.4 illustrates the problem surrounding the development of a new product where the old product has benefited from the learning acquired over a long period. Unless it exhibits a marked improvement, unlikely in a mature industry, its initial unit costs are likely to be higher and to remain so until its cumulative volume reaches V (the point of break-even with the old product). A short-term sacrifice must be accepted if the long-term benefits are to be achieved. The following points can be inferred from the graph:

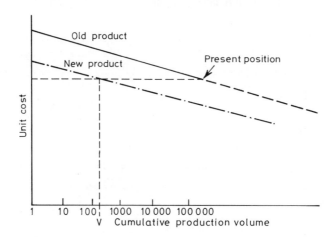

Fig. 6.4 *New product decisions and the experience curve.*

1 The greater the performance improvement in terms of unit cost the sooner it becomes competitive; the attendant risks are also likely to be greater.
2 A comparison of current production costs of the existing product with initial costs for the new will show the latter to a disadvantage; if the experience curve effect is not taken into account the potential of the new product (or process) may be sacrificed.
3 The price that can be charged may not be affected; in this case there will be an initial loss of profit, which forms part of the investment. Indeed it may be necessary to charge a lower price in order to build up market penetration.
4 A high-volume manufacturer has an inherent advantage over a low-volume manufacturer, indicating the benefits to be derived from a high market share; strategically it is usually preferable to have a large share of a relatively small market than a low share of a large market.

Make or buy
There is always the choice of whether to manufacture or to buy from a supplier. The strategic considerations, including the pros and cons of vertical integration, have been discussed in Chapter 2. Such decisions can also be made at the operational level. It would seem that it is financially more desirable to purchase if the price is lower than the company's own manufacturing cost. But what is

Financial implications of engineering decisions

the manufacturing cost? Similar arguments apply here to those discussed earlier. The price to be paid must be less than the actual real savings per unit in manufacture.

There are a number of other factors which need to be taken into account:

1. Cost is not the prime consideration, but the profit contribution is. Thus it may be advantageous to purchase if the company's own facilities can be diverted to other more profitable activities.
2. The purchase price is a variable cost whereas the investment in one's own facilities has a large fixed element. Thus if the market is volatile or the economic climate uncertain it may be preferable to disengage from manufacture. In this way the market uncertainties can be minimised.
3. The subcontractor may have the benefits of a larger production volume. If this is not reflected in his current price quotations it may leave room for renegotiation.
4. Control of production and quality is not under one's own control. This may appear a powerful argument but may on closer examination be shown to be a myth.

6.5 Value added

The concept of value added is often used as a measure for evaluating the performance of a manufacturing company. It can be argued that it is an appropriate measure, for it represents the wealth created by the operations of the firm. It is defined as:

$$\text{value added} = \text{turnover} - \text{materials and services purchased}$$

or

$$\text{value added} = \text{pay and salaries} + \text{interest and dividends} + \text{depreciation} + \text{tax} + \text{retained profit}$$

The importance of value added is that it focuses attention on the productivity of the engineering activities and emphasises wealth creation. It also enables comparison to be made with other business by deriving the ratios of value added per employee, and value added per £1 of capital employed.

Examination of the first of the definitions indicates that the value added is the difference between turnover, which is an output measure, and the materials and services, which are inputs. As such it is a measure of the technical content of the product. There is, for example, relatively little difference between the materials and services purchased to manufacture a Porsche or a cheaper mass produced car. But there is a much greater difference in the price, reflecting a much higher value added. This is not solely a consequence of the engineering input, since value added can also reflect the ability of marketing to create the right image.

The second definition directs attention to the trade-off between wages and depreciation, a measure of the capital intensity of the company. In general the aim should be to minimise the total of these two elements, taking into account any additional interest payments which might be incurred.

Value added is not a criterion which can be used in isolation. It is only one of the calculations, albeit an important one, in relating financial considerations to technical policies.

6.6 Conclusions

The purpose of this chapter has been to focus attention on some of the most important financial aspects of the engineer's corporate role. It is apparent that this cannot be performed without an understanding of how to use accounting data in his decision-making.

A number of difficulties have been noted. Throughout emphasis has been placed upon the contribution to the company's profit performance, which is not always synonymous with cost reduction. The engineer must attempt to reconcile what from time to time will appear to be conflicting objectives, particularly since cost reduction is often easier to appreciate and to quantify.

The need to choose the appropriate figures in assessing engineering decisions has been stressed. This can create misunderstanding when using the accounting figures routinely prepared by the organisation. The accountant is concerned with consistency from one period to another, and has to conform to accounting conventions and internal financial policies. The engineer, on the other hand, needs to select those figures which are relevant to his decision at the time it is taken. He must always ask whether at the end of the day the company is better or worse off as a result of his decisions and select the appropriate figures to quantify them. The different purposes for which the accountant and the engineer require figures must be appreciated.

Figures always possess an aura of accuracy. This is often spurious since they contain a large number of judgements. This is particularly true in relation to engineering decisions. Decisions relate to the future, which cannot be known in advance. Budget and investment decisions are always based on estimates, and as such incorporate judgements relating to what must be to a large extent unknown at the time they are made. The assumptions underlying these judgments must be examined and questioned.

Few engineering decisions can be made solely on the basis of financial calculations. They involve many factors that are hard to quantify even though they have a financial influence at the end of the day. For example, the costs of quality improvements are relatively easy to express in financial terms, but not so the benefits.

Central to all this is the relationship which exists between the individual accountant and the engineer. This must be founded on mutual understanding and a degree of knowledge. The engineer cannot be expected to have an intimate knowledge of accountancy; that is not his job. But he must know sufficient to ask the right questions of the accountant and to interpret the answers.

Some years ago the author was invited to visit a multinational manufacturing company. He was met and conducted round the R & D laboratory and the factory, his guide explaining the technology in a convincing manner. At the end of the tour the guide introduced himself as the chief accountant. The company had an outstanding productivity record. It was apparent that the engineer in that company took an active interest in encouraging the involvement of managers from other functions. The message is clear. If the accountant appears remote and obstructive the engineer should take the initiative to encourage his interest. It is not sufficient to be passive. In addition, the engineer should make the effort to understand the role of the accountant. Bridges cannot be built without bridge builders.

Chapter 7
The management of technical change and technology transfer

At a time of rapid technological progress fundamental changes to the structure and organisation of a company are often essential in order to exploit it fully. Significant changes are also frequently required at lower levels in an organisation when more limited innovations are proposed. It might be assumed that a competent managerial team, following the logic of the concepts described in the preceding chapters, would ensure that the company at all levels would be receptive to new ideas and able to implement them. This assumption would, in general, be incorrect. There are always considerable barriers to change of any kind even when the need for it may be recognised.

The past 30 years have been marked by the rapid growth of formal business education and the intellectualisation of many managerial activities. Management today is regarded as a profession with a body of knowledge and a wide range of techniques to aid in planning and decision-making. Notwithstanding this, it is only part of the overall managerial function. It is not sufficient to recognise the need; actions have to be taken to introduce the necessary changes. Furthermore, on frequent occasions the need itself will not be widely appreciated. This is particularly true where the change is stimulated by technology. The potential may be understood by the engineer but he is likely to be unaware of or insensitive to the wider company impact of the technical innovations he advocates.

In this chapter we shall examine the processes by which innovations are introduced into companies, discuss the nature and reasons for the barriers to change, and suggest means whereby these barriers can be overcome. Nothing contained in the chapter detracts from what has been described previously. The concepts and procedures of strategy formulation, new product development, managerial control systems and so forth are essential but they only provide a framework within which individuals can work effectively. It is individuals who are responsible for effecting the changes. Although what is written here is directed towards the engineer and technical change, the same considerations apply elsewhere in the organisation.

The management of technical change and technology transfer

It will be noted that the chapter title includes the terms technology transfer and technical change. Technology transfer is the process whereby technological knowledge is captured by the organisation (an input) or transmitted outside the company, largely embodied in products or through licensing (an output). Technical change comprises the managerial and organisational processes necessary for the acceptance and implementation of new technology.

7.1 The process of innovation

Research into the success and failure of innovation highlights the critical role of individuals in ensuring success. In technological innovation that individual will normally be an engineer. The terms used to describe him include: technical or project champion; technological entrepreneur; and (more recently) intrapreneur. He may or may not be the originator of the idea.

The champion is typically a man of action who is undeterred by opposition and is often prepared to jeopardise his own career in the pursuit of his objective. He is necessary because most human beings oppose change when it affects them individually even though most people would give verbal support to the need for change in the abstract. There are, of course, some organisations which are highly receptive to new ideas, but they are the exceptions and are more likely to be found in the newer than in the more traditional engineering companies. A feature of such companies is that they recognise the importance of the champion and adopt organisational structures and procedures to encourage and support him.

The more radical the innovation, the more important is the role of the champion because the degree of organisational resistance is likely to be greater. In Chapter 4 (see Fig. 4.4b) it was noted that many successful innovations are brought about by a process which might be regarded as premature development without the foundation of theoretical knowledge. This is usually a consequence of the intervention of a champion who seizes an idea and is determined to make it work even though major uncertainties remain; these he will ignore or brush aside.

The champion is single minded and uncomfortable to work with. He can also be extremely dangerous if the project he espouses contains some fundamental weakness in relation to either the technology or its wider impact on the organisation. If the wrong person is given a free hand without some form of constraint, this is a recipe for corporate disaster. Thus he has to work within a system of formal evaluation. Systems for evaluation were discussed earlier. They can be regarded as a filter to ensure that poor projects are rejected and good projects approved. However, if the filter is too fine it can be used to reject the majority of proposals. Furthermore, when used in this way it will result in the diminution of creative proposals and the frustration of the champions who may well leave the company. Thus the key to any innovation success must be striking the right balance between the encouragement of the innovator and the application of formal techniques for selection and evaluation. Many sophisticated companies have introduced so many rigorous managerial systems that they have destroyed the internal climate conducive to the introduction of change. This danger has been recognised more widely in recent years.

The elements necessary for the process of innovation can be seen to be:

An innovative environment   This is a reflection of the corporate culture and its attitude to risk and innovation. It is set by top management and there is little the individual engineer can do to change it. A business which is managed entirely by a top-down philosophy is unlikely to provide the right environment, for a large number of the changes must be effected from the bottom up. Top management cannot abdicate its ultimate responsibility, but the successful company at a time of rapid change is likely to combine communication and the basis for actions both top down and bottom up.

A source of ideas   Creativity has been discussed earlier, when it was noted that the generation of ideas is highly dependent upon the environment both corporate and within the engineering departments.

The technical champion   Without the support of an individual fully committed to the successful implementation of the change, it is likely to fail in spite of the inherent merits of the idea itself.

An evaluation system   The choice of the system is of less importance than the way in which it is used. If it is applied too rigidly, innovation will be killed and change will not occur. If it is too slack, projects with low potential may be approved.

If any one of these elements is missing, successful innovation is unlikely to occur. The key managerial judgement is establishing the correct balance between the need for a formal system and the recognition of the importance of the human element.

7.2 Human barriers to change

In all organisations any proposal which involves change will meet resistance. This is as true of the engineering function as of other areas of corporate activity. Every person exhibits it, including you, the reader. The only difference is that others oppose change because they are narrow minded and unenlightened, whereas we like to think that our own opposition is based upon sound reasoning. But is this really the case? How do you react when someone proposes that you change your working practices or technology? Almost always the initial response will be negative. There may or may not be good reasons for this reaction. But the same arguments apply equally to those who oppose your proposals. Let us now examine some of the reasons why this occurs in all organisations.

The management of technical change and technology transfer

Lack of top management commitment
The importance of the role of top management in establishing the climate for change has been stressed earlier in relation to attitudes to innovation and risk. There are, however, two other aspects which can have an important bearing on their willingness to accept technically stimulated change:

A short-term orientation Many technological innovations have a long lead time before they yield a profit to the company. Opposition may be a consequence of what we might like to refer to as shortsightedness. But this is not always the case. Uncertainties in the business environment or financial constraints may impose genuine restrictions on the freedom of top management to make investments for the long term, however much they desire them.

Ignorance of technology In many companies few of the directors have a technical background. Their ignorance of technical developments diminishes their appreciation of the business potential of technological innovation and enhances their perceived appreciation of the associated risks.

Modifying these attitudes is not easy but it should be regarded as part of the engineer's corporate role. With respect to the time scale of innovation it is often possible to modify an engineering proposal so that it can satisfy both short- and long-term criteria. Overcoming problems of a lack of engineering understanding is often a matter of good communication between the engineer and the non-engineer. The latter is not primarily interested in the elegance of an engineering proposal. He is concerned with its implications for the business. These must be identified and communicated in his language, clearly and with the avoidance of technical jargon.

The generation gap
Creative ideas are more likely to be generated by the younger and less senior engineers than those who are older. There is evidence to indicate that younger people are inherently more creative. To this must be added the fact that in a fast moving technology they are more likely to be aware of the latest developments.

In general, however, they will have little experience of the practical problems of transforming an idea into an end product and of the wider organisational implications of their proposals. There is often an inability to communicate effectively with their superiors or to express themselves clearly on paper. In some cases they may appear to be brash and impatient, even intolerant of their superiors. In other cases they may lack the confidence to pursue their ideas in the face of opposition.

It might be argued that these considerations are irrelevant to the merit of the ideas, which should be examined objectively. This may be so, but in an organisation consisting of people, all of whom have their own faults, the reality of a non-rational response must be recognised. All engineers should assess their communication skills and endeavour to improve them. The young engineer must consider carefully how best to convince others of the merits of his ideas; he must 'market' them. He may be convinced that he is right but this is of little value if he cannot convince others. The more senior engineer must recognise that many potentially good ideas will come from the more junior members of his staff. He must create a climate within his department which encourages the flow of ideas from lower levels, and explain his reasons for rejecting them if this is necessary.

Premature critical evaluation
The engineer is trained to be rigorous in his analysis. Because of this he is likely to exhibit more strongly than most people a human trait of being critical. Most people when presented with a new idea will show more enthusiasm for seeking its faults than its merits. However, it is extremely likely that most ideas for change will suffer from some weaknesses. This does not imply that there are necessarily insurmountable difficulties; problems there will be, but these must be regarded as challenges to be solved.

The study of creativity shows that one of the greatest barriers lies with this tendency to evaluate ideas prematurely. Indeed many of the creative problem-solving techniques stress the need to suspend judgement in order to allow the idea to mature. If this does not happen, potentially good ideas are likely to be rejected.

There must, of course, be a stage when every proposal is analysed in detail. However, if the negative aspects are over-stressed on the initial presentation it may well be rejected. This indeed may give some intellectual satisfaction to the senior manager who 'sits in judgement' and shows his intellectual superiority through his ability to destroy a proposal on the grounds of identifying flaws overlooked by the person putting it forward. This trait, although undesirable, is almost universal in human beings. The reader should examine his own reaction when new ideas are suggested, particularly if it affects his own performance. The response must be to hold back this initial negative reaction and to seek the merits of the idea before looking for the weaknesses, and when discussing the weaknesses to do so positively to explore methods by which they may be overcome.

The most important impact of premature evaluation is the effect it has on the proposer. He may feel he has been made 'to look a fool'. It may destroy his confidence in his own ability. It will almost certainly discourage his willingness to submit any future ideas he may have. In this way many potentially good ideas will be lost to the business.

Fear of change and risk aversion
The uncertainties associated with any proposal to introduce change carry with them a risk, not only to the business, but also to the individuals whose activities are identified with the change programme. All individuals suffer to some extent from a fear of the unknown. Many are averse to taking risks if they can be avoided.

The risk aversion can reflect the personality of the individuals affected by the proposed change. They are comfortable with the status quo; past experience has armed them with tools for dealing with emergencies; there is little than can happen for which there are not well tried solutions. In some cases the fear of the unknown extends beyond the aversion to risk and can create stress and illness amongst those who feel they are unable to cope with the new situation.

Uncertainty implies that there is a possibility of failure. The failure may reflect some aspect of the change programme that could not have been foreseen. It may, however, also occur as a result of deficiencies in the way it has been managed. It might be hoped that the differences between these two causes of failure are clearly understood within the organisation. This is often not the case, largely because it is usually difficult to distinguish between them and partly because companies rarely conduct a post audit to establish the causes of failure. Thus the people involved believe that their careers are at risk by involvement with the unknown and that they may be penalised for shortcomings over which they had no control. This may well deter a potential project champion.

If there is a genuine risk to those involved in a change to accepted practice there is no logical reason why it should be accepted unless there is a commensurate potential personal reward. In large companies with formalised salary scales it is often impossible to reward individual success except by promotion which may not always be feasible. Thus there can be a mismatch between the top management which desires innovation and the attitudes of those through whom it must be effected. Although few people will attempt to assess the personal risk/reward ratio for a course of action, most are aware that it exists and are unlikely to undertake a risk which is both personal and corporate unless the anticipated rewards are personal as well as corporate. This is a factor which few businesses address.

For some people the risks associated with change are likely to be real and significant. If a new technology makes a job irrelevant for the future, it is not surprising if those affected oppose its introduction. This is a rational response which is understandable and must be recognised. On some occasions, however, these fears are ill-founded but persist because no managerial attention is directed to allaying them. If a person has a perception of a situation, that perception is real to the individual even where it is imaginary.

This barrier to change is of greatest significance where those who appear threatened are senior managers. Resistance to the many manifestations of information technology often arises because senior and older managers feel threatened by the skills and knowledge possessed by their juniors. This is exacerbated by the latter's use of technical terms or jargon which raises barriers to good comunication. Were this not the case, many of those who feel threatened would realise that the new technology does not present a real threat since the understanding required to do their job is usually not deep.

Self-imposed perceived constraints

Every person has a perception of what he can and cannot do within his corporate responsibility. Rarely are these limits clearly defined. A formal job specification cannot be sufficiently specific to cover all eventualities. It is drawn up to give general guidance on what a person is expected to do. This is often interpreted as implying that he should only act within this set of guidelines and that it defines what he cannot do. Often one hears comments such as: 'It is a good idea but I would not be allowed to do it'; or 'That would not be acceptable in this company.'

But is this really so? Social science research indicates that many of these limits do not apply in reality. They are self-imposed perceived constraints. It is useful for the individual to analyse the basis for these perceptions. Has he tried to introduce the change and been rebuffed? The usual response is that he has not, and if encouraged to do so will find that the objections and opposition he anticipates are imaginary. Business success has often been based upon companies breaking the bounds of what its competitors have traditionally regarded as being outside their expertise. Industrial restructuring has resulted from the challenges of new companies working within a different paradigm. These arguments also apply within the business and at all levels.

The assessment of the strength of opposition to change may lead to an unnecessarily negative attitude. Although these barriers undoubtedly exist they are often weaker than anticipated. One area where this often occurs is in employee resistance to technical change. For example, a research study into the adoption of robots in British industry (Policy Studies Institute, research report 660, 1986) found that 31% of companies expected opposition to their installation from the shop floor and from trade unions; in practice, however, only 2% of robot users actually experienced this difficulty whereas 17% reported that labour relations had actually improved. In this context it is important to draw a distinction between technical change and organisational change although the former often necessitates the latter. This same study indicated that resistance to organisational change, particularly amongst employees, is much greater than to technical change.

Thus in wishing to bring about a technical change or technology stimulated organisational change the engineer should analyse carefully the sources of likely opposition. He should then question whether his assumptions about this opposition are soundly based; often he will find that they do not stand up to this critical examination. Sometimes, of course, he may reveal sources of opposition he had not previously suspected, for the reasons discussed earlier. Frequently, however, he will find that many of his fears are imaginary.

7.3 Evaluating a technical change programme

In considering the impact of a technical change the engineer's thoughts are likely to be confined to the effect within his own sphere of operation. His evaluation of it is likely to be focused on technical performance and an economic analysis of its potential benefits. Where these changes are significant, their repercussions are likely to extend outside his own limited area, sometimes affecting all functions in the company. These are managerial and corporate implications. The analysis of potential technical and economic benefits is focused on tangible aspects which are usually amenable to expression in quantitative measures. His training and natural inclinations prepare him for this form of analysis. However, the opposition arises from people – employees, senior managers in his own function, and managers in other areas of the business. It is these people whom he must convince. Furthermore, the merits of his proposal may not be so apparent to others who may be assessing it in relation to different criteria. In order to overcome these difficulties he must be prepared to:

1 Communicate the benefits of the technical change in terms that others can understand and that are relevant to their needs
2 Be prepared to modify his proposal, either to make it more palatable to those who may be adversely affected by it or to meet wider organisational objectives
3 Develop a strategy for introducing the change which maximises its chance of success.

In this section we shall address the second of these factors, and then in the next section examine the development of a strategy for change implementation. There are a number of questions which the engineer should ask himself.

What are the merits of the idea?
The answer to this question may at first sight appear self-evident. On closer examination, however, other merits may appear, particularly in relation to the concerns of other parts of the organisation. The objectives of an advanced manufacturing system were discussed in Chapter 5, where it was noted that what is regarded as paramount at the corporate level, for example quality or responsiveness to the market, may be different from those considerations uppermost in the mind of the engineer.

The most important impact of the engineer on the business is where he is proposing a major change in the technology of either the product or the process by which it is manufactured. It has already been stressed that with the rapid advance in many technologies the frequency of such changes is increasing, and so is the corporate role of the engineer. It may be taken for granted that such proposals would not be put forward unless the engineer is convinced of the technical merits. Although others may question these merits they are generally not competent to assess them in detail and must rely to a great extent on the engineer's judgement. Their concerns will relate to the impact of the introduction of the new technology upon their own chief interests. In an ideal world these interests would be confined to their assessment of the corporate impact within their own sphere of responsibility. In the real world of people more personal consequences cannot be overlooked, since the technology may enhance or detract from the power or status of senior people in the company who have an influence on the decision-making process.

Thus the merit of a proposed technical change cannot be isolated from the corporate environment. What may appear to be a highly desirable change from a simple techno-economic analysis is not necessarily so when viewed from a wider perspective. What may be appropriate for one company may not be so for another. In order to maximise the chance of success the engineer must understand the business as a whole, its corporate objectives and the organisational constraints. When he communicates with other functions he must consider how they would be affected by the technical change he is proposing and assess its merits in relation to the criteria important to them. It is not intended to give a comprehensive list of these criteria here, but a few will be briefly discussed as illustrations:

Finance The finance function is not interested solely in the ultimate ROI of the investment. The phasing of the project in relation to cash flow, the availability and cost of finance, or the impact on short-term profit projections may be of greater immediate concern.

Marketing There may be a ready market for a product incorporating the new technology, but this does not necessarily mean that it is within an area where the company has marketing expertise. It may, for example, require a different distribution channel. In order to exploit the potential it may be necessary to recruit new people or even come to some arrangement with another company which does have the expertise; the latter is a corporate decision. For example, ICI found it desirable to form a new company, Marlow Foods, in conjunction with RHM to exploit its biotechnology expertise with a company possessing knowledge of the food market.

Personnel Almost inevitably a new technology will have an impact on the numbers and skills of employees, including both professional (particularly in R & D) and managerial staff. Retraining may be required. Solutions to these problems must be found. One Japanese company retrained operational employees to be laboratory assistants, thereby enabling them to run a three-shift system in their laboratories with a significant impact upon product development times. Although such a solution may not be feasible in the European culture, it does illustrate the scope for innovative responses to the impact of new technology.

If the engineer can anticipate these impacts on other members of the organisation he may be able to identify aspects which have positive benefits for them. These can be valuable selling points for the technical change. But he must also address those aspects which will meet resistance, not only to enable him to promote counter-arguments, but also to modify his proposals as far as possible in order to minimise the opposition.

How complex is it?
Many engineers are purists. It is attractive to develop optimum solutions to problems, perhaps introducing a number of technical changes at one time. Sometimes this may be desirable; we have already discussed the undesirability of incremental change in new manufacturing technology. But this is not always the case. It may be possible to modify the proposal so that it is more compatible with existing systems, thereby reducing the degree of organisational change and reducing the opposition from others in the company. Although this might involve some degradation of the anticipated benefits, this may be an acceptable sacrifice if it means that it will tip the balance between support and rejection. The aim must be to achieve a result which is within what is possible in the business in relation to its overall objectives and the attitudes of those most affected. The engineer must avoid the arrogance of believing that his desires are sacrosanct and should not be modified on the assumption that others must modify their practices in order to fit into his. In many cases it will be found that the exercise of creative thinking and the introduction of minor changes in the design can have a major effect upon its acceptability. These changes, incorporated before the proposal is put forward, may be less radical than those which may be necessary after initial opposition has been aroused.

How should it be implemented?
Since every change involves some degree of uncertainty it can be expected that a major change that affects the whole of a product range or manufacturing process will involve a greater risk than one that can be introduced on a smaller scale. These risks can be reduced by:

1   Testing the new technology in a pilot study
2   Implementing it by a step-by-step approach.

It is common practice to explore the market acceptance of a new product by means of a test market, where it is launched into a limited geographical region deemed to be representative of the country as a whole. This is an experimental approach to gain knowledge to reduce the uncertainties and is of much greater value in the reduction of risk than sophisticated risk analysis computations. If the test proves to be a failure then the product can be withdrawn with the minimum of cost. It might be thought that the experimental approach would appeal to engineers since it is common practice in the solution of technical problems. This is often overlooked in the management of technical change; engineers frequently propose major new investments where a less ambitious programme would enable some of the uncertainties to be resolved. Often this is a consequence of the company being a technical follower; major change is required in a short time to meet an immediate crisis resulting from a realisation that the company's products or costs have become uncompetitive in relation to others. By anticipating the long-term need for change there is time to experiment with alternative technical and managerial approaches and remove the greatest uncertainties before a major investment is made. In order to do this it may of course be necessary to modify the engineering design.

A technological innovation may involve the use of a number of new technologies as well as having a variety of impacts on other parts of the company. If all these changes are introduced at one time the risk of problems in several areas occurring simultaneously is high. The ideal situation would be the introduction of one change at a time to permit the concentration of technical and managerial resources on the solution of any problems that may occur. Between the two extremes of incorporating all changes at one time and only one at a time there is often a wide range of alternative programmes. These should be explored to see whether the original proposal can be modified in order to achieve a phasing of change to minimise the impact over a short period. As we have noted earlier this is another example of how the engineer may be able to reduce the organisational resistance through modification of the design or programming of this proposal.

It is important to draw a distinction between the magnitude of change and the rate of change when considering human resistance to it. Individuals need time to come to terms with any need to modify well established practices. Thus their resistance is related to both the size and the speed of change. In this section it has been suggested that the engineer should examine his proposals for technical change in order to devise ways in which both the size of the change occurring at any time and rate of change can be minimised, whilst aiming to achieve his ultimate objectives. This may mean that the planned period for the introduction has to be extended, though in practice it is less likely to suffer from the disruptions accompanying a more ambitious programme. These factors again point to the early initiation of change rather than delaying until it is necessary to embark upon a crash programme.

What is at risk
In Chapter 6 we explored the financial consequences of uncertainty. Uncertainty means that there is a probability, which can be subjectively quantified, of a project failing or falling far short of its planned performance. This does not necessarily imply that all the investment will be lost. It is important to draw a clear distinction between the total investment and that proportion of it which will be sacrificed in the event of failure.

At one extreme we have the failure of a major new product to find a market, for example an aircraft, or a special plant capable of producing only one chemical. In these cases the total investment is at risk. At the other end of the scale is a simple modification to a product or a piece of equipment which can

The management of technical change and technology transfer

be returned to its original state if necessary; in this case the risk is minimal. Most proposals for technical change will fall between these two extremes. Thus in considering a project the engineer should ask:

1   What is the total investment required?
2   How much of this investment is irrecoverable in the event of failure?

Although answers to both these questions are relevant to the finance director, it is the second which may sway his decision. Furthermore, it is an aspect over which the engineer has some control since he should examine ways in which this risk can be reduced by modification of his design.

How easily can it be understood?
Engineers have considerable difficulty in communicating their technical ideas to non-technologists. In presentations they frequently stress technical features which their audience cannot comprehend and which would probably be of little interest to them if they could. Technical terminology, precise in its meaning to the engineer, may be meaningless jargon to others. It is, of course, highly desirable that accountants, marketing men, personnel managers and the top management of an engineering company have some understanding of the products the company makes and how it makes them. In many successful engineering companies they will have a familiarity with engineering matters but they cannot be expected to have a detailed knowledge of highly specialist technical subjects.

Their concern is how an engineering project affects them in terms of strategy, money, markets and people. This is where the engineer needs to focus his attention. To do this he will of necessity have to describe what it is he proposes to do in language that is meaningful to them. However, this part of the communication process is only a small portion of the total message; all too often it is the major focus. It must be remembered that what the engineer is proposing is of far less importance to them than its effect upon the organisation, both internally in the way it affects their vital interests and externally in relation to its profitability.

Summary
In this section we have discussed a number of questions the engineer should address before putting forward a proposal for technical change. The answers to the questions may frequently require him to modify his design or programme in order to minimise possible objections. Once again, the questions are:

1   What are the merits of the idea?
2   How complex is it?
3   How should it be implemented?
4   What is at risk?
5   How easily can it be understood?

7.4 Gaining support for technical change

In the preceding sections it has been noted that the majority of technical changes are initiated within the engineering departments, either R & D or design, and that in many cases they result from the efforts of an individual who may meet resistance from his colleagues or superiors. We have also noted that it may be desirable to modify the proposal in a variety of ways in order to reduce opposition from other functions within the company. In this section we shall examine ways in which the individual might then proceed in order to maximise his chances of gaining approval.

This activity can be likened to marketing except that in this instance it is necessary to market the proposal within the business. Having established the merits of his idea for various interests within the company the engineer should identify the key individuals or opinion formers who are influential in the decision-making process. In most cases this will be the senior managers although this is not invariably the case. These people will fall into two classes. The first is those who would benefit from the proposed technical change. The benefits must be analysed and presented in such a way that they are made explicit; it is not sufficient to assume that they will be self-evident. The second class is those who will be likely to oppose the change for a variety of reasons either genuine or perceived.

In order to maximise the probability of acceptance it is essential to gain the support of those who will benefit from the technical change and make them allies at as early a date as is feasible. This can be done by involving them in the thinking and planning of the project so that they become enthusiastic about it and identify with it. In this way the nucleus of an influential corporate team can be created.

The greatest problem arises from those who are likely to be in opposition. There can be several reasons for this:

1. They may be insufficiently convinced of the merits; in this case imaginative presentation of the project may bring about a change in their views.
2. There may be valid objections in relation to the effects of the change on the part of the business for which they are responsible; this might indicate the need for modification of the proposal to reduce or remove the objections.
3. They may be change averse for a number of reasons which have been described earlier in the chapter. The engineer must use his tact, persuasion and imagination to reduce these fears.
4. They may be adversely affected personally. In the extreme case their skills may be made redundant by the new technology. These are likely to be the major source of opposition.

Frequently technical changes receive setbacks or even fail because of resistance that had not been anticipated, particularly where organisational change is

Table 7.1 *Technical change implementation audit: a systematic analysis*

Persons affected	How affected?		Why?	How might they react?	What response should be made to their reaction?
	Favourably 1 → 5	Unfavourably 1 → 5			

Within the company
1 Top management:
   Corporate strategy
   Personal factors
   etc.
2 Finance:
   Cash flow
   ROI
   Capital budget
   etc.
3 Marketing:
   Segmentation
   Distribution channels
   Expertise
   etc.
4 Production:
   Equipment
   Managerial systems
   Scale of operations
   Skills
   etc.
5 Personnel:
   Managers
   Employees: Numbers
               Skills
   Training
   etc.
6 R & D:
   New technology
   Skills
   etc.

External to the company
1 Competitors
2 Customers: Existing
              New
3 Final consumer
4 Suppliers
5 Sources of finance
6 Shareholders
7 Subcontractors
8 Trade unions
   etc.

168   The management of technical change and technology transfer

involved. In most cases these problems should have been foreseen and plans made to eliminate them or reduce their impact. The reactions of those who will oppose the change must be anticipated and preparations made to counter them. A quick response can meet the challenge before attitudes harden. This response can only be made when it has been thought through in advance.

Table 7.1 shows a draft audit form which can provide the basis for a systematic approach to the analysis of the problem. This audit has proved helpful in a number of technical change programmes. It forces the innovator to assess the impact of the technical change on all those who are likely to be affected both inside and outside the company; to assess the magnitude of the impact; to gain an understanding of why they are likely to oppose or support it; to anticipate how they might react; and to plan to meet that response. This analysis must be carried out before the proposal is formally submitted in order to ensure that the engineer has examined all its ramifications throughout the company and has laid the best possible foundations for ensuring its success. Furthermore, he will have prepared those who are likely to be his supporters.

Many engineers rely entirely upon the technical merits of an idea and are unwilling to identify with the full range of corporate activities of which they are only a part. Some may regard the process of persuasion and lobbying as distasteful. Unfortunately many worthwhile proposals for technical change have failed to gain corporate support through this parochialism. The engineer must play an active part in the total process; if he fails to do so he must bear a major responsibility for the slow rate of technical change in his company.

7.5 Organisational structure for technical change

The purpose of an organisational structure is to provide a framework within which people with different functional responsibilities can work together to further the objectives of the business as a whole. Its main characteristics are to make explicit who reports to whom and the channels of communication. There is a choice of different forms of the organisational structure and the aim must be to select that which is most appropriate to the needs of the business.

When the technology/industry life cycle was discussed it was noted that the needs of different stages of the development of a technology demanded different managerial styles and systems. Furthermore it was seen that it is necessary in a multiproduct company to embrace these differences within the one organisation. In this case it is possible to overcome some of these problems by separating the different types of activity. Even more difficult is the management of technical change within the mature part of the business where the emphasis is placed more upon manufacturing innovation than upon new products.

The characteristics of technical change which the organisational structure must aim to facilitate are:

Responsiveness to new information However detailed the planning for the change programme there will be many occasions for modification in the light of new information and experience. This can lead to confusion and resistance to the change if those affected by it are not fully informed of the details of the modifications and the reasons for them.

Good cross-functional communication Since most technical changes will have a much wider impact throughout the company than normally required in the day-to-day operational conduct of the business, the organisational structure must encourage close contact and communication between the functions at many levels within the hierarchy. In a stable situation formal procedures largely eliminate the need for this close personal contact.

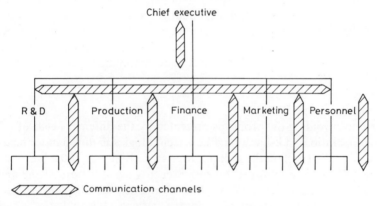

Fig. 7.1 *The hierarchical organisation structure.*

Management related to the needs of the change programme Where a radical technical change is involved the importance of the technical champion has already been noted. The organisational structure must facilitate the exercise of this role. In the absence of a champion it is still likely that a project manager will be appointed. In either case it is necessary to ensure that the person appointed is responsible for the whole project including all the functions involved.

A number of alternative organisational structures will now be examined to see how well they meet these requirements. It must be remembered, however, that the needs of managing technical change are not the only consideration. There will normally be many activities which are unaffected by major change which must be managed effectively where the needs are different. This means that there is usually no ideal solution and that the organisational structure must of necessity be a compromise. It must aim to facilitate the total activity of the business, not hinder it.

The hierarchical organisation
Figure 7.1 shows the hierarchical organisation typical of a large number of engineering companies. It is characterised by a high degree of separation of

functions. An individual reports through the hierarchy to the head of his function. The flow of information with other functions is up through the hierarchy and across by communication between departmental heads. In a stable organisation where change is infrequent this structure is appropriate. The demands from one function to another can be formalised and infrequent. For example, marketing can inform production of their requirements for the next period and the detailed schedule can be prepared from this information based upon experience and planning documents. Engineering budgets can be prepared based upon known costs for a year in advance. There is little need for detailed communication between the functions.

It can be seen that while this organisation structure is based upon the division of specialist knowledge where each person has a well defined role, it does not meet the requirements for technical change noted earlier. All co-ordination between functions is done at senior management levels, where those concerned cannot have the detailed knowledge of all the cross-functional implications of a change. This inevitably leads to inadequate communication and a low rate of response.

New venture groups
One way of avoiding the problems presented by the different needs of running a stable operation and developing new technologies is to separate them completely, managing each in the style appropriate to that part of the business. New venture companies established by entrepreneurs are a feature of the development of new technologies. What these companies usually lack is adequate financial resources and a depth of business experience. It would appear desirable to have a combination of the entrepreneurial characteristics of the venture company and the financial and managerial resources of a large parent.

The fact that large companies often fail to exploit the new technologies can be attributed to their inability to understand the different needs of a rapidly expanding new technology and those of the current business. The application of the rigid managerial systems of the existing company within the traditional hierarchical structure inhibits the creativity and enterprise which are essential. This is not inevitable. In Japan, for example, there are relatively few new venture companies and the new technologies are developed within the large businesses.

In recent years this problem has been appreciated by an increasing number of large companies, particularly in the USA. The 3M company, for example, has for many years based its expansion upon technological innovation and an organisational structure designed to support the entrepreneur. In its purest form the new ventures are established on green field sites where the entrepreneurial chief executive is given a high degree of autonomy. The argument in favour of this is that the company would not have supported him unless it believed he was the best person to bring the project to a successful conclusion. However, he does have the facility to draw upon the parent company's financial and specialist resources. There is, of course, a great temptation for the top management to interfere when they observe something of which they disapprove; although well intentioned, such actions usually hinder rather than assist the project.

One of the reasons for the traditional approach is that the manufacturing operations are seen as the centre of the company and activities related to innovation as being on the periphery – perhaps even as an irritant. When the rate of technical advance is great this structure can be turned inside out; the nucleus of the company is a collection of new venture groups and manufacture is carried out on dispersed sites not necessarily owned by the parent company. For companies in these industries it is technical innovation, not production, that provides their competitive advantage.

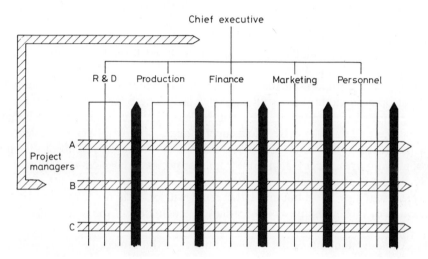

Fig. 7.2 *The matrix organisation.*

The establishment of an autonomous venture group is not always appropriate where its functioning depends upon the sharing of expensive facilities, say a pilot plant, which it would not be economical to duplicate. In such situations difficulties over priorities of access cannot be avoided. Top management must establish the priorities and concern itself with the management of the interfaces between the activities, ensuring as far as is possible that the interests of the venture manager, who may be relatively weak in terms of organisational power, are safeguarded.

Lip service is often given to the formation of venture groups or similarly titled groups within large companies. All too frequently they are constrained by so many organisational restrictions that they are venture groups in name only, with little freedom to make their own decisions. Often they do not include all the business functions necessary to launch the new project as a genuine addition to the company's existing activities.

Matrix organisation

Some form of matrix organisation is being adopted by an increasing number of companies (see Fig. 7.2). In this structure the project manager has overall cross-functional responsibiity for the successful conclusion of the project, thereby satisfying the three conditions necessary for technical change discussed earlier. It will be seen, however, that an individual member of the team has two bosses. In relation to the technical competence of his work and career development he is responsible to his functional head, but the management of his activities is the role of the project manager. It can also be seen that the project manager has no direct authority over the functional heads in respect of priorities and the allocation of staff. It defies two of the classical rules of management – 'one man one boss' and 'no responsibility without authority'.

It must be appreciated that in a complex organisation there must be a matrix of communication whether or not it is recognised formally. Inevitably there must be differences of priority and ambiguity which must be resolved. One person must be responsible for the overall management of the project. Differences of opinion cannot be avoided, particularly where the technical change is being applied to an ongoing operation such as in manufacture. In the absence of a project manager to drive it to a conclusion the change is likely to be delayed or frustrated by interfunctional disagreements at the highest level. For the only people who are concerned with the whole project as a business activity are the chief executive, who will be too busy to give it his detailed attention, and the project manager. Without the latter there will be little drive to complete it. Because of the ambiguities of the situation it demands of the project manager:

1 An understanding of the business impact of the project
2 A commitment to achieving a timely and economic conclusion
3 The managerial skills to ensure its accomplishment
4 A high level of diplomatic ability to gain the co-operation of those functional heads whose managerial support he requires.

There are many forms of project management. The matrix organisation implies an almost equal division of power between the project managers and the functional heads. The venture group organisation centres the technical change almost entirely upon the project with little or no influence from the functional heads; indeed the project manager can be likened to the chief executive of his own small company. At the other extreme we have people entitled project managers who operate wholly within one function. In some companies the title is applied to people who are little more than progress chasers with no power to influence the achievement of the business objectives of the company. To be effective, however, some structure approximating to a matrix or venture organisation is essential for the efficient management of technical change. Difficulties cannot be avoided but the growing adoption of matrix organisations in engineering companies indicates that these difficulties can be overcome.

7.6 Management for technology transfer

In earlier chapters the need to capture new technology has been discussed together with a description of the need for monitoring the total technological environment and the role of the gatekeeper. We shall now explore how the business can adapt to ensure that this process is achieved with the maximum effectiveness. Frequently this will involve major corporate decisions such as the relocation of the R & D department.

The need to manage the process of technology capture and technology transfer is becoming of increasing importance as all companies are exposed to increasing international competition. Thus, although it is important to seek sources of new technology within the company's home country, because they are usually the easiest to identify, this is not sufficient if technological and business opportunities are not to be missed. The most important of these trends are:

1   The widening international technological base
2   The escalating cost of R & D
3   Rapid international communications
4   The growth of international markets
5   The growth of international production
6   The growth of international R & D.

The widening international technological base
Any country today is the source of only a small proportion of the new technological knowledge being generated. For example, UK R & D expenditure is about 5% of that of the OECD. Thus even if a company has good contacts within the UK research community this will only enable it to have access to a limited extent to all the potential sources of information.

Even more important is the fact that nations develop their own competencies. No one nation, including the USA or Japan, can expect to achieve excellence over the whole range of scientific and technological endeavour. This has the effect of increasing the degree of specialisation between nations. This is most marked in the science-based industries where the critical developments for the future may lie in one department of one foreign university; indeed in some cases the expertise may depend upon one man or a small team of researchers.

The escalating cost of R & D
In many industries the cost of all aspects of the technological development process is rising rapidly – basic research, applied research, design and product

development. This applies equally at the national level and to the individual company. If the developments are to be undertaken some form of co-operative effort is essential where the pooling of resources makes possible programmes which would otherwise be beyond the capacity of any of the partners working alone. Within the EEC, for example, a range of programmes have been initiated (ESPRIT, EUREKA, BRITE) in areas regarded as essential for the future of European competitiveness. A common failure of many such programmes and also work carried out in nationally funded research institutes is that they have little impact on industry; the technology transfer does not occur. Industry may often state that there are good commercial reasons for not embracing the new technology generated by these programmes. Although this may be partially true there is a great deal of research of potential industrial value which is not exploited. It is the responsibility of the engineer to identify, assess and adapt where necessary this resource which could be of value to his company; no one else is competent to do it.

Another growing trend is the collaboration of countries, often across national boundaries, in product development. The European Airbus is an example where the resources of individual nations could not support the development of aircraft to challenge the dominant position of Boeing. In aero engines this co-operation involves not only European but Japanese and American companies. Where full collaborative projects may be deemed inappropriate, technology transfer between companies by means of technical exchange agreements are often made. Thus Siemens (Germany) has agreements with Philips (Netherlands), Bull (France) and ICL (UK), and Intel (US) has an exchange agreement with Toshiba (Japan). It is worth noting that the companies which appreciate the need for such exchanges are amongst the strongest in their industry, not the weakest. There are, of course, many potential dangers in exchanging commercial secrets with a possible competitor, but the fact that an increasing number of major companies consider it essential for their future success indicates that this form of co-operation should be considered by all companies.

The problems of working with competitors are of less importance in the early stages of applied research, where the need is to establish the basic technology upon which products will be based at a later date. Although not yet widespread it is likely that this form of 'pre-competitive research' will become more common in the future.

Rapid international communications
The speed of communications means that information and people can be transferred internationally in a short time. This does not necessarily mean that the technology transfer will occur. Communications only facilitate the process where there is both a willing transmitter and a willing receiver. Personal contact is still the most effective way of enabling technology transfer but this can only be brought about where the individuals are prepared to travel and the company supports it. The gatekeeper can only function when there is an acceptance of this

role within the company. This view is supported by research into internationally successful firms, which showed that they:

Support external and international contacts
Support professional visits
Encourage continuous education
Encourage conference attendance
Support publication of results.

The growth of international markets
The internationalisation of the markets for engineering products is a process with a long history. In the past it was largely a matter of exports from the major industrialised nations to less developed parts of the world. In recent years two trends have had a significant impact on the process of technology transfer through the export of products:

1   The growth of the number of countries capable of manufacturing engineering products, e.g. Taiwan and South Korea
2   The rapid expansion of trade between the developed nations, e.g. in the motor car.

Thus every company is now facing international competition in its home market but also has a greater opportunity of selling in overseas markets provided the design and cost of its products are competitive. Nevertheless, the acceptability of a particular design or product is still influenced by a number of local factors. One of the most difficult of these is often provided by national standards, which vary greatly between different countries. It can be expensive to modify a product designed to meet one set of national standards to satisfy the requirements of another. These considerations should be incorporated in the initial design in co-operation with the marketing department; there is no merit in adding cost to a product in order to meet the requirements of a market which the company has no intention of entering.

It must not be forgotten that the market is the aggregate of buying decisions by individuals whose needs and tastes vary from one market to another. In the UK, for example, the market for many domestic appliances is approaching saturation, with the notable exception of the dishwasher whose numbers are low compared with other developed nations; the dishwasher manufacturer must decide whether this provides an opportunity because of the size of the potential market, or the reverse because of the resistance of the market to acquiring this type of product. The electric kettle has been a feature of the British home for many years; only recently has it been supplemented by the boiling jug, which was the standard appliance in Australia 25 years ago. It would appear possible that an enterprising company could have transferred this technology from Australia much sooner than actually happened, provided of course that it would have been acceptable at that time. In France a saucepan is still widely used for boiling water. These examples indicate the variety and complexity of the different markets in spite of the growth in international trade. This must be reflected in the design of the product and requires close integration of technical and market knowledge within the company.

The growth of international production
Many companies now manufacture in several countries. There are a number of reasons for this:

Manufacturing cost
Availability of specialist skills
Financial inducements
Legal constraints and tariff barriers
Distribution costs.

It should be noted that whereas overseas production was initially introduced to satisfy the market of the host nation it is now common for the dispersed manufacturing units to supply world markets. Alternatively they may supply only components which are incorporated into products assembled elsewhere. The strategy adopted by a company must relate to its individual circumstances and the characteristics of the industry. The overall objective must, however, be the supply of products meeting the needs of the individual market at the lowest cost. Inevitably overseas production adds organisational complexity and communication problems. These factors must be considered carefully, but it is a strategy which has been adopted by many of the most successful international companies, e.g. IBM. In some cases European companies have only been able to survive by siting some of their manufacturing capacity in another country which has a market advantage in respect of one of the factors listed above, e.g. electronic components in the Far East.

In the UK motor industry two recent developments illustrate the impact of some of the factors. Nissan established its British factory in order to overcome EEC import restrictions, but the siting decisions were influenced by government financial inducements. These were also factors in the Rover/Honda co-operation. However, in the latter case the partnership also embodied the expertise of Rover in designing the car body and the financial resources of Honda R & D in mechanical design.

The growth of international R & D
There is a worldwide shortage of competent engineers. Furthermore, as we noted earlier, technical excellence is dispersed. These two factors are reflected in the siting of R & D partially or wholly in other countries. Thus we find European companies establishing laboratories in India where there is an availability of cheap highly qualified engineers. For similar reasons a number of

The management of technical change and technology transfer 177

American pharmaceutical companies site some of their R & D in the UK, as does IBM.

A characteristic of growth in high-technology industries is the establishment of a critical mass of competence in a particular location. One example of this is Silicon Valley, which owes its origin to the research carried out in the local universities. This led to the establishment of a rapidly growing number of small entrepreneurial high-technology companies, often by former academics who maintained their close links with the universities. Some of these companies are now large. This concentration of expertise provided a pool of knowledge which other international competitors found it hard to match. In order to gain access to this expertise a number of European and Japanese companies have now established their own laboratories and sometimes manufacture microelectronics in Silicon Valley.

The establishment of an overseas laboratory is a major corporate decision reflecting a commitment to technology. More limited in scope is identifying key people and persuading them to work in the company's own facilities. Although the term 'brain drain' is widely used to describe the transfer of European engineers to North America, this is but one example of a growing international trend which is not confined to movement from the less developed to the more developed nations. Only engineers within the company are able to identify the key workers in their own field; if their co-operation is considered to be of great importance to the company a variety of alternatives must be examined. These could include a technical exchange or consultancy agreement, offering them employment or providing them with laboratory resources in their own country.

7.7 A corporate framework for technology transfer

A number of factors have been discussed in the previous section. These can now be summarised and presented in a systematic approach to the management of technology transfer in the form of a check list.

Which are the critical areas of science or technology?

1 *To provide opportunities for development current products or processes* It must be remembered that some of the most significant commercial advantages can be gained by the incorporation of new technologies in the development of improvements to existing types of product, e.g. new materials and electronics in the motor car.
2 *Critical for the future* The identification of such areas can be aided by the use of technology forecasting.

Where does the expertise lie?

1 *Within the company* An audit of the technological abilities of people within the business may reveal sources of knowledge or expertise not being fully

exploited currently, or individuals who could easily acquire the required expertise with limited training.
2 *In other companies* These could be in the:
Same industry, in the same country
Same industry in another country
Another industry in the same country
Another industry in another country.
3 *In universities, research institutes, consultancies etc.*
4 *In one or a few individuals*
5 *In suppliers or customers*

What could we do? – Strategic

1 *Take over or merger* with a company which possesses the desired technological ability, marketing expertise or financial strength. The purpose of this type of strategy is to acquire the essential resource (which the company lacks) that is needed for commercialisation of the technology.
2 *Collaboration, partnerships or joint ventures* with companies in the same or complementary industries in order to:
Pool research funds, e.g. pre-competitive research
Spread development expense
Acquire markets or marketing expertise
Add to or relocate production.
3 *Relocate R & D* nationally or internationally:
To be near universities with the required research base (e.g. Hewlett-Packard has laboratories near Edinburgh and Stuttgart universities; Pharmacia moved its laboratories to be near Uppsala University)
To be near a critical technological mass (e.g. Silicon Valley; Route 128 in Boston; Cambridge Science Park)
To gain access to a source of trained technological personnel.
4 *The purchase of technology*:
Knowledge through licensing or contract R & D
Products, by purchasing components or subassemblies from the international leaders for incorporation in one's own product.
5 *The sale of technology* This is the reverse of the purchase of technology; whereas the former was concerned with technology transfer into the company, this is concerned with the capitalisation of the business's R & D by sale.
6 *Technical exchange agreements*

What could we do? – Operational

1 *Develop new expertise in-house* Although this is the traditional approach and is often desirable, it suffers from the problem that it may not be possible

to recruit the best people initially and it usually takes several years to build an effective team in a new field of technology.
2. *The formal monitoring of technological developments* This involves gaining access to international sources of information. Although the English language is widely used throughout the world nowadays, it is not universal. Much valuable information is still published only in German, Japanese or Russian – a useful source which is largely neglected.
3. *Head hunt* This is the recruitment internationally of leaders in the field who bring with them knowledge which it would be extremely difficult for the company to acquire in any other way.
4. *Develop academic networks* This not only helps to keep the company in contact with the leading research but also assists in identifying those institutions where the most important work is being undertaken. The funding of sponsored research in those universities or the association of faculty members through consultancy should also be considered.
5. *Identify gatekeepers and use them* The importance of the gatekeeper role has already been stressed. Although this section is concerned with a systematic approach to technology transfer it can be little more than a framework or check list. In all respects it is dependent upon the knowledge, insights, enthusiasm and entrepreneurship of individual engineers. This will not happen, however, if management, including engineering managers, do not consciously create the corporate environment within which this process can occur.
6. *Get out and about* Focusing attention on the gatekeeper should not be allowed to engender an attitude where technology transfer is regarded as solely the responsibility of a few individuals. It is every engineer's responsibility. His sources of information are:
Conference attendance, meetings of local engineering groups, participation in events organised by his professional institution
Reading of technical and technology management literature
Visits to customers and suppliers
Technical and management course in the UK and overseas.

In conclusion it can be seen that there are three vehicles for technology transfer – *knowledge, people* and *products*. All three must be used in a systematic approach. In a world of rapid international technological growth this is an activity which cannot be left to chance; it must be managed. Parochialism and the mistaken belief that a company can by itself meet the technological challenges of the future can only lead to a deteriorating competitive position. In many companies this fact is insufficiently understood; in such cases the engineer has an important role to play in educating and bringing about change in the organisation; it cannot be left to others. But in all cases he is the vital link to the outside world of technology.

7.8 Conclusion

This chapter has explored the process of technical change, the human barriers to change, organisational structures and the management of technology transfer. It has stressed the role of the individual engineer in all these processes. There are three important elements that have been considered:

(*a*) The corporate environment and organisational stucture
(*b*) The intellectualisation of the problems and a systematic approach to them
(*c*) The commitment to the introduction of technical change and the engineer's corporate role in bringing it about.

In every organisation the engineer will encounter difficulties, be they in the attitudes of those with whom he has to work or in the organisational structure. There is no such thing as the ideal company peopled by an enlightened group of managers and workers. In spite of the difficulties which have to be overcome the engineer must not be discouraged in his attempts to effect technical change. He must apply his intellect to the problems and above all be active in initiating and implementing change whatever the obstacles he may encounter. Only thus can he make an effective contribution to the business.

Chapter 8
The engineer: career progression and development needs

The first seven chapters of this book have examined business from a variety of viewpoints highlighting the contribution of the engineer. A number of common themes have emerged. It might appear to the reader that this paints a picture of needs which any one person is unlikely to be able to fulfil. Furthermore, in order to achieve them within the limited space of an engineering career there must inevitably be some sacrifice of engineering expertise if attention is also to be paid to wider business issues. It may seem that the rapid advances in technology alone are sufficiently great that it is more than a full time job to keep up with them and to introduce them into the engineering products or processes with which he is directly involved without having the additional task of acquiring knowledge of the wider business environment.

It would be facile to assume that there is a simple solution to this dilemma. Each individual must find answers that are appropriate to his own abilities, inclinations and career aspirations. This can only be done, however, if there is first a recognition that a problem exists. Armed with this knowledge he can then identify others who are able to provide the expertise he lacks himself. It must not be forgotten that the individual engineer does not stand alone; he is part of a team and a member of a number of networks of information which he must use to the full in order to complement his own expertise. As his career progresses with the assumption of wider responsibilities he must reassess his information needs and develop his own systems to acquire it.

This chapter will attempt to pull together the threads of the arguments presented earlier and place them within the context of the business related decisions the engineer has to take and of his career development.

8.1 The changing nature of business

It is useful to start by summarising some of the most important considerations which have been constant themes throughout this book:

The increasing corporate role of the engineer

This arises from the changing nature of the market and the development of new technologies. In an advanced society the key determinant of many buying decisions is no longer price but a combination of one or more of the product attributes required by sophisticated and increasingly affluent purchasers. For many industries it has been noted that the most profitable companies are those which have concentrated on high value-added products focused upon a particular segment of the market, often relatively small in terms of volume. The major contributor to this added value is technology. In many organisations future prosperity also depends upon technology-based business diversification. Thus in both situations it is technology which provides the most significant contribution to the company's competitive advantage. It is, therefore, essential that the engineer plays a significant role in the formulation of the company's strategy for the future. He alone can identify and interpret the threats and opportunities presented by technological developments.

It might be argued that this is a role for the most senior technical managers and has little significance for the majority of engineers. However, if these managers are to discharge this responsibility to the full they cannot hope to acquire the necessary understanding of the totality of the considerations relevant to the strategic application of technology only when appointed to a senior position. They must prepare themselves for it in advance. It must also be recognised that the complexity of modern technology is such that the senior technical manager is highly dependent upon the flow of information to him from more junior engineers. They also have a valuable part to play in assessing the business implications of the new technological knowledge they possess. In order to do this they must develop a business orientation even though they will not need to have a detailed grasp of all the strategic complexities.

The strategic importance of technology varies with time. Although it is always an important business resource there are periods in a company's history when it may not be the critical factor for achieving a competitive advantage. However most engineering companies are now competing in an environment of increasing technical advance which is likely to remain the major contributor to business growth for the forseeable future.

The global nature of the business environment

The generation of technological knowledge, the location of manufacturing, the nature of the competition and the markets for all engineering companies must today be analysed within a global context. This does not necessarily imply that the companies must become multinationals. However, even for those organisations whose operations and markets are confined to a single nation the importance of foreign sources of information and the potential threat from overseas competition cannot be ignored.

This has a number of implications for the engineer. He must widen his catchment area for new technology to sources throughout the world. He must be aware of the applications of new technology in the products of all companies even though they may not be competitors currently; they could be in the future. In recent years many cultural, legal, economic and logistic barriers which afforded protection in the past have disappeared. Indeed the greater the degree of protection, the greater the vulnerability when these barriers are removed. For example, the opening of British Telecom procurement to international suppliers is but one illustration where companies are suddenly exposed to competition from which they have previously been sheltered.

Another effect of this globalisation is the increasing number of engineers who spend part of their careers abroad and the frequency of overseas visits for those working at home. In many cases this will necessitate the learning of one or more foreign languages. Some engineers welcome this; others resent it as an unwelcome disruption of a settled way of life with undesirable consequences for the careers of their spouses and the education of their children. But this is a fact of life which cannot be ignored. Because of this widening of the engineer's experience owing to the exposure to a broader range of stimuli it is highly likely that those engineers who have travelled most widely will develop those attributes which lead to career advancement.

The pace of change
Many aspects of business are experiencing an accelerating rate of change. Some of these reflect a long-term trend, particularly in respect of technology and the widespread effects of information and communications technology. In many cases the long-term changes can be identified and forecasts of sufficient accuracy made for most business decisions. Others, however, exhibit rapid fluctuations over a short period reflecting an increased degree of turbulence in many elements of the business environment. Currency exchange rates and commodity prices, particularly energy, are areas where it is almost impossible to predict with accuracy even for relatively short periods ahead.

All these factors lead to a greater degree of uncertainty than existed in the past. In some technological fields the pace of product substitution is increasing rapidly. For example, the traditional gramophone record had a life of several decades, with the introduction of LP and EP records the only major innovation. Then tape cassettes, the compact disc and digital audio tape were developed, each following more quickly than its predecessor. There is uncertainty about which technical alternative will become dominant and also about the rate of substitution. Technical change has occurred at a time of growing international competition, rapidly changing consumer purchasing decisions and a volatile economic climate. Many other industries have experienced similar conditions. Unfortunately for the majority of the technologies the development times for the new products have not decreased; indeed in some they have increased.

The combination of all these influences makes companies much more vulnerable to incorrect decisions. No longer is there time to wait to see what the competition does and then hope to recover the consequences of a wrong decision by reacting swiftly; in most cases it will be too late. Although the situation will be different for each company, some general conclusions can be drawn:

1 A long-term orientation is needed in order to identify significant trends, particularly in technology; this implies that greater attention should be paid to all forms of forecasting including technology forecasting.
2 A greater emphasis must be placed upon the quality of decision-making since the consequences of a poor decision can be extremely serious; this calls for a greater attention to be paid to the intellectual contribution to the decision-making process using those management techniques that can be of assistance.
3 The shortening of product lives in many industries adds urgency to the minimisation of development times. Good project planning and project management are consequently becoming even more essential for business success than they were in the past.
4 Wherever possible business and engineering plans must be made flexible so that they can respond rapidly to unforeseen events; this may require the adoption of a sub-optimum solution.

However, it is not only at the strategic level that the pace of change is accelerating. The introduction of CAD/CAM, CIM and just-in-time provide the opportunity for both engineering design and manufacture to respond rapidly to short-term changes in the market in respect of both product design and the volume required. Although the system provides the capability its potential can only be realised through the actions of engineering managers.

It is not difficult for the engineer to understand what is happening; it is much more difficult for him to develop the ability to cope with the rapidity of these changes. He must accept that change is now continuous rather than an infrequent discontinuity in his activities. This involves the constant critical evaluation of current technologies, product designs, manufacturing processes and procedures, and his contribution to the overall objectives of the business. In order to do this he must be willing to acquire new technological knowledge and skills, to develop his decision-making ability and, above all, to acquire a sensitivity to the changing needs of his business. The ability to make quick decisions is, however, of little value unless he can translate them into prompt action. For it is not sufficient to reach the correct decisions; they must be brought to fruition before the competition. The individual engineer cannot do this by himslf since his decisions are implemented by others in whom he must inculcate the same degree of urgency. Thus he must regard himself as an agent for change and the motivator of those upon whom he depends for success.

Closer integration of business functions
The pace of change and the need for greater organisational responsiveness and

The engineer, career progression and development needs 185

flexibility demand the rapid transmission of an increasing volume of information between the functions of a business. Information technology provides the mechanisms whereby this can be achieved, but it is only an enabling system. It can only be effective when those using it have a high degree of mutual understanding. No system can by itself replace the need for face-to-face dialogue and understanding of the problems confronting others in different functions. This becomes increasingly important with the rate of change.

Thus all engineers who have contact with other departments must have a knowledge of and a sensitivity to their main concerns and priorities. As we have seen earlier the number of engineers who are involved in cross-functional communication increases with the pace of change. This has led to new forms of organisational structures (see Chapter 7). A common feature of these structures is the project manager, who has responsibilities extending beyond engineering and crossing functional boundaries. The term 'project manager' can be confusing since it has traditionally been used in engineering in a narrower sense. Project management in the wider sense described in this book is being adopted by an increasing number of engineering companies. But project management requires good project managers. Many organisations report that the lack of good project managers is one of their most serious weaknesses. Although project managers are not necessarily engineers the nature of the work usually dictates that they will be. The four key attributes of the project manager are:

1 The knowledge, ability and motivation to take project decisions in the light of the overall business objectives
2 The ability to combine engineering, marketing, manufacturing and financial criteria in making project decisions
3 The ability to select the appropriate data, analyse them and take effective decisions
4 Personal drive and interpersonal skills to ensure that his decisions are implemented.

It will be seen that the first two of these attributes demand a knowledge extending beyond engineering to the business as a whole. There is a critical shortage of engineers who possess this knowledge. It is a weakness that is becoming more widely recognised. To some extent engineering courses are now incorporating some elements of business and management education but there is relatively little that can be done within the time constraints of a traditional three-year first degree course. In-company training can also make a valuable contribution to rectifying this deficiency, but the volume of this training is often inadequate. The conclusions must be that this is of such importance to the engineer's career development and his contribution to his company that he must take the initiative himself to develop his knowledge of the work of other fractions and the business as a whole in the absence of any formal training provided by his employer.

The integration of technologies
A feature of the products of most engineering companies is the widening range of technologies they incorporate, many of them science based. This is not confined to microelectronics, which is all-pervasive, but also includes new materials and such hybrid or convergent technologies as biosensors and protein engineering. The business exploitation of these advances depends firstly on the recognition of the potential available from the combinations and secondly the ability to acquire and combine the necessary skills. Most engineers, however, are trained primarily in one discipline.

It must be accepted that the identification of the opportunities offered by technologies outside the traditional activities of the company must lie within the technical department. But who within the technical department is equipped to do it? This is a difficult question to answer since the rapid rate of advance in all technologies is such that it is a difficult enough task to keep up to date in only one of them. However, at the identification stage it is not necessary to possess a deep knowledge of other potentially useful technologies. What is required is:

1. A scanning and monitoring system to become aware of developments in the whole field of science and technology
2. The ability to identify their key attributes
3. The creative ability to see how these attributes might enhance the performance of current products or lead to the development of new products.

In most companies this process is left to chance. However, it would appear desirable to place it upon a more formal basis where it is thought to be important for the future of the company. In order to succeed there must be individuals with a wide-ranging scientific curiosity who can be identified and then used. This process can be seen as analogous to that of the technical gatekeeper where the company can only build on the motivations which exist already. One can also see an analogy with the general manager whose knowledge embraces all functional activities although he still retains the greater depth of specialist knowledge stemming from his early training and employment. Thus one can see emerging the 'general technologist'. It might also be argued that this role should not be left to just a few individuals, but should be of concern to all engineers.

8.2 Engineering and business decisions: some differences

Engineering decision-making always involves some element of compromise between conflicting considerations. However, the engineer's training is largely science based; this implies a search for absolute truths based upon rational analysis. In the world of practical engineering the realities of the situation force upon him a more pragmatic approach. Nevertheless there remains the natural desire to seek for the 'right' solution, although there may be some doubt about what constitutes this right solution.

With this background the engineer may consider that many of the decisions made in business seem irrational. In some instances he may, of course, be right but he must appreciate that the complexities of business decision-making are far greater than those involved in most engineering decisions. There are so many uncertainties that the judgement of the decision-maker must be an essential element. This does not mean that rational analysis should not be applied wherever appropriate. But underlying this analysis there will be many assumptions which reflect the judgement of the manager. Frequently there will be a number of alternative solutions from which the final decision must be selected. In such circumstances it is impossible to think of the 'right' solution; it is necessary to seek the 'best' or even the 'least bad'.

A number of the conflicts in business decision-making have emerged in earlier discussions. These will now be reiterated for the sake of completeness and some additional considerations introduced.

Survival versus growth

Most businesses aspire to grow in the long term, but in order to do so they must survive. A large investment for the future may starve the company of the financial resources which may be needed to overcome more immediate crises. However, an over-cautious management may be reluctant to commit sufficient resources for the longer-term business development through a desire to have adequate funds to meet any conceivable eventuality, however low the probability. The role of the financial institutions and institutional shareholders is an important factor to consider in this context. It is often stated that their concern with company short-term profit performance is a major deterrent to investing for the future. This may or may not be the case but many finance directors believe it to be so, thereby influencing their investment decisions.

The balance between short- and long-term investment is one of the most important strategic decisions. In arriving at it the board must be mindful of outside factors such as the attitude of the financial markets as well as their own personal judgements. The most usual manifestation of this, as far as the engineer is concerned, is the reluctance to fund long-term engineering developments for which the arguments may appear to be irrefutable on the basis of the more limited criteria that he will consider in his analysis. In the wider context the strength of these arguments may be diluted.

Quantitative versus qualitative information

The information upon which any decision is based is partly quantitative and partly qualitative. The quantitative information can be further divided into objective or factual data such as the volume of last year's sales, and subjective or probabilistic data such as the expected return on investment of a new project which, whilst expressed in quantitative terms, reflects the judgements of those making the estimates. The application of mathematical techniques such as risk analysis and decision trees enables a great deal of qualitative information to be operated on as if it were quantitative factual data. This enables decisions to be taken which are consistent with the manager's judgement exercised on each of a number of related factors. Knowledge of these techniques should be a part of all engineering managers' skill requirements. However, in most decisions there will remain some elements which cannot be expressed meaningfully in quantitative terms.

The engineer is primarily concerned with factual information and with measurement. His education rightly stresses the need to quantify wherever possible. This is essential for most of the criteria associated with technical decision-making. Thus he is suspicious of subjective judgements and is inclined to dismiss them or underweight them in his decision-making. Business decisions, however, usually involve far more factors, many of which cannot meaningfully be expressed in quantitative terms. The information on which they are based is often woolly and ambiguous. In some circumstances it might be possible to acquire more detailed information but the cost of obtaining it or the time required to gather it may be inappropriate in relation to the nature of the decision and the time when it must be made.

It can be seen that these considerations can lead to conflict, misunderstanding and poor communication between the engineer and others in the organisation for a number of reasons:

1 The engineer may disregard or undervalue the importance of factors which cannot be expressed quantitatively, although they may be the key criteria for the success of a proposal.
2 Non-engineers may place too much emphasis on their judgement and intuition and be reluctant to quantify and think in probabilistic terms.
3 The engineer may be reluctant to take a decision until he has collected and analysed all the information he considers essential to remove uncertainties and to ensure that he has the best solution.
4 The non-engineer may be more interested in reaching a decision quickly on the view that an acceptable answer now is of more value than a better one at a later date.

These problems will become more acute as the engineer rises through the management hierarchy. Whilst retaining his concern for rigour he must accommodate himself to the realities of an environment in which decisions must be made with inadequate information, where unquantifiable factors can be of central importance and many of the uncertainties do not lend themselves to quantitative analysis.

Importance versus priority
Important decisions are those on which the long-term future of the organisation

is dependent. Usually they will place great demands on the future of the organisation, but because of the relatively long time scales involved it might be argued that it is possible to defer a decision. There are also other claims for resources which may need to be satisfied urgently. For example, an investment to improve the efficiency of a machine tool with a limited life might still offer a higher ROI than a major re-equipment, but only if the work is initiated immediately. If the decision is delayed the working life may be sufficiently curtailed to invalidate it. In such cases it may be decided that the more favourable financial return of this proposal should gain it priority.

Although this may appear to be the same as the long-term versus short-term argument the reasons behind it are not the same. Priority may be awarded because of urgency if the rewards are to be gained rather than a reluctance to take the long-term view. However, it can be seen that in practice the results may be similar in that a series of investments made on this basis can lead to incremental developments, thereby reducing or eliminating the resources required to meet long-term objectives. The converse can also occur where profitable short-term investments are sacrificed in the interests of the important long-term projects. As with all aspects of management the answer must lie with striking a balance; where the appropriate balance lies must be a matter for managerial judgement.

Engineers are likely to have a number of proposals for priority projects which they believe to be outstanding in relation to the criteria they apply – technical performance and ROI. Often these proposals are rejected in favour of others which are considered more important in relation to the objectives of the business. This can cause considerable frustration unless the reasons for rejection are communicated and understood.

There is also another class of projects which it may be difficult to justify on the normal criteria but where sanction cannot be deferred indefinitely. The re-equipment of a canteen or the building of a new office block may be difficult to justify in comparison with an important long-term engineering project. They do not contribute directly to the profitability of the company. To many engineers such expenditure may be regarded as unimportant, particularly if it is made at the expense of technical alternatives. Yet there will be occasions when such expenditures must be accorded priority since they cannot be disregarded indefinitely.

Entrepreneurship versus management systems
This aspect has been discussed at length in earlier chapters, where it was seen that change is usually brought about by the entrepreneurial action of individuals. Formal managerial systems on the other hand are essential for the efficient management of stable situations. As the pace of change accelerates the role of the internal entrepreneur is becoming of increasing importance and is recognised in the organisational structures of a growing number of companies. This does not invalidate the need for formal systems but it does mean that their

design must be sufficiently flexible and their operation sensitive to the needs of innovation and the entrepreneur.

Striking the right balance between these two requirements is one of the most critical areas for innovative engineering management. In many companies the formal systems which have been introduced to improve managerial control and reduce cost have had an inhibiting effect upon the introduction of change, which depends to a great extent upon the creativity and entrepreneurship of individual engineers or engineering managers.

It has been seen that some companies have been able to maintain an innovative momentum through the development of novel organisational structures which encourage entrepreneurial activity whilst maintaining the formal disciplines necessary to avoid anarchy. It was stressed earlier that there is no right managerial style – only one that is appropriate for the activity being undertaken. Thus in a typical engineering company, where there must be a combination of tight management of the existing activities and an encouragement of innovation, there must also be a diversity of management styles. This problem becomes most acute where both activities take place in the same location. What are the implications of this for the engineer?

1 Most importantly he must recognise that these differences occur and understand the reasons for them.
2 He must accept that he has a personal responsibility for introducing change and should not allow himself to be deterred by the opposition he is likely to face.
3 If he has been working in an innovative environment he must accept the necessity for the introduction of more formal control systems as his project or the business matures.
4 He must analyse his own aptitudes and ensure that he works for an organisation where they are appropriate. Not all engineers are innovators. Non-innovators are unlikely to succeed in an innovative environment; the converse is equally true. Although it might be argued that the company should match people to their aptitudes this will frequently not happen in practice. The engineer himself should not be passive about something upon which his own career is so reliant. If there is a mismatch he should seek a transfer within the company or seek employment elsewhere.

Individual, departmental and corporate values and objectives

The successful company is that in which everyone is pulling in the same direction. This may appear a self-evident truth, but few organisations achieve it. It is not infrequent to hear engineers criticise the company for which they work or the more immediate management of their department. Of course there will always be some managerial deficiencies in all companies since no combination of human beings can hope to be perfect. However, in many companies there is low morale, which can have a serious impact upon the effective operation of the business. There are a number of reasons why this can occur.

Every company has its ethos or value system although it is rarely spelt out explicitly. If there is a rejection of the corporate value system by the individual engineer then he really has no alternative but to seek employment elsewhere, for he is unlikely to have much influence in changing it. This may happen only in rare cases but corporate social and environmental issues can on occasions raise ethical problems for the individual.

A more important cause of difficulties can arise where the aims and values of a department or function are not consistent with those of the business as a whole. Often this is a reflection of the professional concerns within a department. In R & D, for example, there may be a desire to explore new technologies which are not accepted as appropriate elsewhere in the organisation, perhaps because there is no agreed R & D strategy. In marketing the aim may be to maximise sales volume, whereas in manufacturing it is to minimise unit costs. On the basis of the considerations discussed in earlier chapters one might conclude that this is a consequence of poor objective setting. However, it must be recognised that however desirable explicit objectives may be they can be of lesser importance to a group of individuals than what they see as their own corporate interests. For example, engineers are concerned with the concept of technical excellence. It forms an important part of their training and is something from which they gain personal satisfaction. They may recognise the need to 'design down to a price' as a business requirement but find it difficult to accept it in their daily decision-making. At the extremes they may allow their professional desires to dominate and design products ill matched to the market, or accept the corporate needs but lose professional satisfaction. The arguments presented in this book support the view that the corporate objectives must be the prime concern. Nevertheless they cannot be achieved in practice if the price is the loss of motivation and perhaps the departure of their best engineers. There must be some degree of compromise.

The formal processes advocated in this book suggest a hierarchy of objectives transmitted through the organisation by means of a series of strategies. This integrated approach is often lacking. The marketing objective may be set as the maximisation of sales volume and the manufacturing objective as the minimisation of unit costs, although there will be many occasions when these two sets of objectives are mutually exclusive. It was seen earlier that in the introduction of advanced manufacturing technology the corporate objectives for its introduction – responsiveness to the market or product quality – may differ from the economic or technological objectives of those who implement its introduction.

Thus although the need for close integration of activities across functions is essential it can be seen that there are a number of influences within the company which can frustrate it. These stem from the clash of individual, professional and departmental values and objectives at all levels. The imposition of a formal system for establishing objectives and strategies to bind together the disparate parts of the organisation, although vital, only provides a framework for decision-making. In practice these needs may be modified, perhaps subconsciously, by the personal inclinations of those involved and by their professional values. In themselves these factors should not be regarded as weaknesses; indeed, the reverse should be the case if they can be harnessed to the needs of the organisation. The organisation can benefit from an understanding of the motivations of the engineer but he in turn must identify with the objectives of his company. A strong engineering department is a corporate asset provided its strength is not used to frustrate the corporate strategy. Thus there must be trade-offs and compromises between the individual, the department and the business as a whole if the latter is to prosper. For the engineer cannot subordinate his own interests to his company completely; nor should he allow them to dominate his behaviour.

Risk versus security
Since there is always uncertainty there will always be risk. This applies not only at the corporate level but also to engineering design and the engineer's career decisions. From the discussion of the changing nature of business it can be seen that the degree of uncertainty associated with the business environment is growing. This applies equally to engineering decisions and the engineer's career decisions.

The linking of the words 'uncertainty' and 'risk' implies that they describe an undesirable state of affairs. On the other hand successful long-term investments, inherently more uncertain, are likely to yield the highest financial returns. Thus one has the alternative linkage of 'risk' with 'opportunity'. The particular connotation attached to either of the linkages gives an indication of the individual's attitude to risk.

Unfortunately most companies do not make explicit their attitude to risk. Their managerial decisions often give the impression that they are risk averse when an examination of the business decisions they take shows that this is not really the case. But a company which is prepared to take risks in the hope of gaining high rewards may frustrate its desires by penalising those within the organisation who take risks and fail through no fault of their own. The risk propensity of a company is a reflection of the corporate culture and there can be no 'right' level of risk, although it must be weighed against the potential benefits. It should, however, be consistent throughout the company. The engineer may appreciate this, but in the final analysis his decisions are likely to reflect the risk/reward relationship as he perceives it to affect him personally through his understanding of the actions taken by the company.

Although the main purpose of this book may appear to be the role of the engineer within one company, the reader's concern is with his own career, which is likely to involve working within a number of different companies. No longer can he contemplate career security within one organisation. Thus uncertainty and risk are as pertinent to him as they are to a company. Although absolute security cannot be guaranteed, some career decisions will involve a greater degree of risk than others. If he is risk averse he may gain more security but he is also likely to forgo the opportunity of significant advancement. This is a personal decision but it should not be taken without a serious attempt to analyse his own risk propensity.

In general two broad approaches to uncertainty have been identified. One of these is forecasting, where an analysis of trends and present knowledge can assist in gaining a view of the future. With increasing environmental turbulence the analysis of trends has proven less dependable for assessing the longer-term future, and techniques which explore alternative futures (e.g. scenarios) are gaining in popularity. The other approach to uncertainty is to build in flexibility so that the response can be varied as the future unfolds. These approaches can also be applied to the engineer in preparing himself for his future career. Some conclusions can be drawn:

1   An attempt should be made to forecast those engineering disciplines for which the demand is likely to grow for the foreseeable future. Gaining an expertise in these can be of value to all engineers but is likely to be particularly appealing to those who are risk averse.
2   Acquiring knowledge of other technologies will enable him to move between industries. A considerable number of senior technical managers are currently working in industries where the basic discipline is different from that in which they were initially trained. This is possible because the scientific method is a constant factor and the depth of technical knowledge required diminishes as one moves into more senior positions.
3   Moving between companies, perhaps in different countries, develops an adaptability and experience which opens up a wider range of potential employers.
4   Transferring between functions broadens the understanding of the interfunctional implications of technology and the knowledge of how a business as a whole operates.

Whilst much of the above has always applied, the current trends in business make them increasingly important for the future. At all times in his career the engineer will have to develop himself through both on-the-job and formal training, if he is to gain the ability to exploit the opportunities offered by a rapidly changing business and engineering environment.

Literacy versus numeracy
The problems of managerial communication have been referred to on several occasions. This is of particular significance when it occurs between the engineer and the non-technologist. The use of technical jargon, for instance, has been cited as a major barrier to the effective communication of technical information to those who do not understand it. The inability of non-engineers to understand

quantitative information has also been mentioned. Technologists often consider them to be innumerate.

On the other hand many non-engineers maintain that technologists are relatively illiterate and are unable to express themselves succinctly and to write good reports. This may be an overstatement but the complaint is so widespread that one must accept that there is an element of truth in it. Most managerial communication is either verbal or written. If the communicator has a poor command of language he is unlikely to be able to present his arguments persuasively irrespective of their intrinsic merits. The good manager must be both literate and numerate. Although the engineer may rightly complain about the inability of others to comprehend quantitatively expressed information, he should also ensure that he develops his own literacy skills.

Much of the informational exchange in business involves advocacy. For it is not sufficient to be confident in one's own mind that a particular proposal is justified. Nor is it adequate to present the information in a bland fashion, leaving the recipient to analyse it. The onus must be on the provider of the information to give it in a well presented and easily understood form, with the keys issues for the receiver highlighted. It may be assumed that the engineer wishes to gain support for his proposals; to do this he must market them and persuade the receiver of their validity. This is an active not a passive role.

Cash flow versus profitability
The often conflicting demands of cash flow and profitability have been discussed in Chapter 6. Although cash flow constraints result from the short-term financial position and profitability for investments is only realised in the longer term, this conflict may not result from a lack of a long-term orientation. It might be argued that if the proposals are sufficiently attractive the company should raise additional capital to fund them. However, this may not always been desirable for purely financial reasons, e.g. the current level of interest rates or the impact upon the share price.

Summary
In this section a number of areas where business management has to resolve conflicting criteria in decision-making have been considered. In particular it has been shown how the training, professional orientation and motivation of engineers may lead to a different weighting of the criteria relevant to decision-making which involves both engineering and wider business considerations. Although there is always danger in describing simplistic stereotypes the engineer is likely to stress growth, long-term investments, future profitability and quantitative criteria and to underweight the constraints of survival, cash flow and qualitative criteria. In addition there are also the needs of innovation *vis-à-vis* formal management systems, possible conflicts between individual, departmental and business objectives, and attitudes to risk.

The engineer: career progression and development needs

Judgement must always play a major part in business decision-making owing to the wide range and nature of the criteria involved. There will rarely be unanimity of view since each person involved in the decision-making process is likely to make different assumptions and assessments of the importance of individual criteria. However, for good decisions all factors relevant must be considered and the judgements made explicit. The more detailed information on which these judgements are based comes from a variety of sources including engineering.

It has been seen that communication problems can occur because the engineer may not:

1. Have a clear indication of the business objective he is addressing, which may not necessarily accord with his perceptions of his own personal, professional or departmental objectives
2. Understand or be sensitive to the complexity of the factors which may have to be taken into account
3. Give sufficient weight to some of those criteria which are of importance to the business decision-maker
4. Use language which is easily understood by others
5. Express himself, both by speech and writing, clearly and persuasively
6. Be sufficiently receptive to the views of others because he does not understand their specialist language, understand their organisational objectives or accept the criteria of importance to them.

These considerations become more important as the engineer rises in seniority. To some extent they explain the lack of influence the engineer has in many companies' strategic management. For it must be stressed that the necessary expertise has to be developed throughout his career.

Finally it should not be thought that the cause of poor communication lies wholly with the engineer; far from it. There is equal need for the non-technologist to develop his skills in communicating with the engineer.

In many ways this discussion is at the root of what this book is about. The thesis is that the engineer must play an active part in the affairs of his business. In order to do this he must have an understanding of the needs of that business as a whole and be able to make the most effective engineering contribution to meeting those needs. This demands knowledge, understanding and communications skills. The alternative is to regard himself solely as a technical specialist to be used by others. This passive approach is unlikely to be advantageous to either the business or the individual engineer. It might be argued that a reluctance to assume this wider role is a major contribution to the ineffective use of technology by many companies and to the low status which many engineers claim is accredited them.

8.3 The engineer's career development

The engineer is initially employed in business primarily because of the knowledge and skills he acquired during his academic education. These will be technical skills relating to one academic engineering discipline with some knowledge of associated disciplines. Only a very few highly qualified specialists will serve their whole career within that specialisation. Within a few years of first entering industry the great majority of engineers will be to some extent engaged in management. Towards the end of their careers a few of them will be discharging the responsibilities of a company director.

On entering industry the career paths described above would appear to most engineers to be the future choices facing them. In practice, however, a considerable number depart from this pattern and find employment in other functions of the business. For example many of the staff of marketing departments in engineering companies have engineering qualifications. Others will find employment outside the engineering industry in such activities as consulting and venture capital management.

What follows is directed towards the requirements of those whose career pattern is similar to that described in the first paragraph. For those who depart from this pattern the acquisition of managerial and business skills will be even more important. There are four main strands of expertise which are required:

1 The acquisition and application of technical knowledge
2 The management of technical activities including decision-taking and the supervision of others
3 Interaction with other functions in the business
4 Interaction with the external business and technical environments.

The relative importance of these elements will vary as the engineer's career develops. For most engineers this will not be a steady evolution and there are likely to be major changes in emphasis from time to time, usually associated with career advancement.

An engineer's career can be categorised as falling into five distinct phases in each of which a different emphasis of his skills and personal attributes will be required. In terms of technical knowledge the initial years of his career require depth of knowledge over a relatively narrow field; these are followed by a widening of his technological base, which of necessity results in less depth in any one field. The need for managerial and business skills increases with periodic leaps, following promotion which heralds his advancement to the next phase of his career. The phases have been entitled specialist, junior engineering manager, section leader, project manager and engineering director in the descriptions that follow. These should be regarded as generic titles, since those used in practice vary between companies.

Phase 1: specialist
The emphasis in this book has been upon management and business. But the resource that has to be managed is technical knowledge, which resides in people. Good management can do little in the absence of engineering skills. The engineering graduate on initial employment possesses this knowledge; it is almost the only asset he brings to his employer. In many cases this knowledge will be more up to date than that of longer-serving employees, particularly in those technologies which are advancing rapidly. This knowledge is primarily theoretical, and he lacks the practical knowhow which is unique to the industry or even to a particular company. For very few does the specialist skill remain his main contribution to the company beyond his first few years.

During this period the engineer is given little discretion over his work. This is likely to be part of a larger project or a portion of a project. His work at this time is closely supervised and its scope limited. It will inevitably require attention to detail and may involve a great deal of routine. As he acquires practical knowhow he is likely to be given more autonomy over his work, although it will remain at a relatively low level. Certain aspects of this role are found to be frustrating to many engineers entering industry for the first time with high aspirations and accustomed to the freedom of an educational institution.

Phase 2: junior engineering manager
When the engineer has acquired practical knowhow and gained the confidence of his superiors, the scope of his work is enlarged. He may be made responsible for a small activity or part of a project. In some cases he will be assigned other staff to work with him on what becomes an activity with which his name is identified. There is enlarged scope for his own creativity and initiative and he is likely to be judged more by his results than his input. With a reduction of detailed supervision he gains greater control of the allocation of his own time. His views begin to carry some weight.

The use of the description 'manager' is slightly misleading, for apart from some degree of supervision he is not significantly involved in managing people. However, it is appropriate in the sense that he is managing his own activities to an increased degree and is taking technical decisions albeit of a limited nature.

Phase 3: section leader
The section leader is responsible for a substantially larger part of the company's activity or a project and will have a number of subordinates reporting to him. This is the first opportunity to manage in the wider sense. The major changes in his career that will occur are:

1 His technical activities will cover some disciplines in which he has little knowledge.
2 Because of this he must learn to delegate and rely upon his subordinates but also develop the ability to assess the key issues whereby he can control their

activities. He is learning to manage without complete personal knowledge.
3 He must accept that the functioning of his section is his responsibility for which he is held accountable.
4 He has to take an active role in planning and controlling the activities of other people.
5 He must develop the managerial skills of communicating and motivating others to implement his decisions, which are no longer solely technical in nature.

The section leader has become a manager and must develop a wide range of skills for which he has probably received little preparation or training. At the same time he is still actively involved in the engineering, although to a lesser extent than previously. In the absence of training many section leaders spend too much time on what they know best, namely the engineering, at the expense of their role as manager. This can lead to two forms of frustration. The managerial problems which he feels inadequate to solve detract from the successful accomplishment of the work by which he is judged. Secondly, there is a sense of failure in that he feels he is losing touch technically because others in his section know more than he does.

This phase marks the first major step from engineer to manager. The problems can become most acute in R & D where the technologist's job satisfaction has stemmed primarily from his ability to solve technical problems. The addition of managerial responsibilities may be regarded as unwelcome distraction from technical activities to which he has devoted his whole adult life. It is difficult for many to accept that management is a profession for which skills must be acquired by education and training. Frequently he has received little formal, or indeed informal, preparation for this major step in his career. It is all too easy for the good engineer to be promoted into a position where he becomes a poor manager. Nor does it follow that the good technologist possesses the personal attributes to become a good manager.

A managerial competence cannot be developed quickly. It is a profession with a large and increasing body of knowledge. The engineer must appreciate this and prepare himself for it from the time he enters industry. The nature of the knowledge required expands as he progresses. At the section leader level the main requirements relate to managing people – team building, motivation and interpersonal skills.

Many companies do have extensive training schemes of which the engineer should make maximum use. However, he should recognise that he has a personal investment in preparing for his own career progression. Few engineers read the technical management literature; even fewer read the general management literature and the ever increasing number of management journals. This applies equally to senior technical managers who devote time to keeping abreast of the technical literature in their own specialist field. Of course the technical manager is still an engineer and cannot and should not abandon his engineering.

But he is also a manager and cannot afford to neglect his managerial development upon which his success is so highly dependent. This applies to all technical managers, but it is at the stage of promotion to section leader that the transition occurs.

Phase 4: project manager
The role of the project manager has been discussed on several occasions. It is at this level that the engineer becomes involved to a considerable extent with other parts of the organisation and finds himself managing a wider range of interface problems. He is fully accountable for what may be a significant portion of the company's resources and yet he is unlikely to control the allocation of those resources to the project. Thus he is in a situation where he must exercise his diplomatic and negotiating skills in order to obtain the resources he requires and to resolve differences of opinion with those other functions with whom he has to work.

Because he is likely to manage more people than the section leader his man management expertise becomes of increasing importance, but so does his need for a knowledge and understanding of the business as a whole. The extent of this knowledge depends upon the nature of the project. Some financial and accounting knowledge is essential since most of the justification for his claims for resources must be made in relation to the financial merits of alternative choices. In most cases he also needs some market knowledge.

It has already been noted that he requires many of the characteristics of the entrepreneur in that he must have a single-minded commitment to his project and the ability to achieve results. Thus he is a man of action, although he is unlikely to succeed without the intellectual attributes of analysis, clear thinking and decision-making. The nature of the role might indicate that it would lead to conflict within the organisation. This is rarely likely to lead to success, which depends upon the performance of the project within the total portfolio of the company's activities.

The role of project manager can be a critical step in the engineer's career. If he fails he is unlikely to obtain the leadership of another project. However, if he succeeds he might expect larger projects in the future and eventual promotion.

It must be repeated that project management is becoming increasingly more popular, there is a universal shortage of good project managers, and it is a role for which all engineers who aspire to managerial progression should prepare themselves.

Phase 5: engineering director
The engineering director is the most senior technologist in the company, acting in the formal capacity of head of the technical activities. He is usually a member of the board of directors. In some companies there may be other members of the board with a technical background but they will have no direct responsibilities

200 The engineer: career progression and development needs

in relation to the engineering function; in other companies he may be the only technologist on the board.

His role is to:

1. Ensure the efficient management of the company's engineering activities
2. Communicate the corporate objectives and strategies to the engineering departments
3. Play the major part in developing the technical strategies consistent with the corporate strategy
4. Contribute to the formulation of corporate objectives, strategy and policies
5. Ensure the contribution of technology in all areas where it has an impact on the corporate objectives, strategy and policies
6. Represent the interest of the technical departments and their staff at board level
7. Maintain close contact with the external technical environment and ensure that technological advances of potential value to the company are identified.

It can be seen that he must divide his time between two main activities, namely his leadership of the engineering function and his membership of the board. In the former capacity he can draw upon the expertise of a large staff to whom he can delegate many of his responsibilities whilst still remaining in overall control

Table 8.1 *Competences required at each phase of the engineer's career progression*

Career phase	Technical competence			Non-technical competence		
	Single discipline	Multi-discipline	Technical environment	Man management	Other functions: finance, marketing etc.	Business environment
Specialist	x x x x x	x	—	—	x	—
Junior technical manager	x x x	x	x	x	x x	x
Section leader	x x	x x	x x	x x x	x x	x x
Project manager	x	x x x x	x x x	x x x x x	x x x x	x x x
Engineering director	x	x x x x	x x x x x	x x	x x x x x	x x x x x

and fulfilling his role as leader. However, on the board he has to rely solely upon his own technological and business knowledge and competence. He is unlikely to make the best contribution to the company if he considers himself to be there solely as a specialist and is so regarded by his colleagues. He is no longer a manager; he is a businessman. This is a transition as significant as that earlier in his career when he made the step from engineering specialist to manager. In many companies the integration of the engineering director as a full member of the board does not take place. There are many reasons for this, but engineers must accept some of the blame in that they may be unwilling or have not prepared themselves adequately for this wider role.

In order to fulfil both his roles he will find it necessary to maintain a large number of external contacts. These may also take the form of membership of government committees, engineering institutions and industry bodies of various kinds. He will probably find it necessary to travel widely. These activities are essential in order that he can gain membership of high-level informal networks, from which he can obtain insights into the technical and business environments relevant in establishing the threats to and opportunities for the technical role in the company. These periods of absence make it essential that he fully delegates the responsibility for routine decisions within the engineering departments.

A distinction has been drawn earlier between the management of technical activities in the company and the management of technology as a business resource. It is this second activity to which he needs to devote his attention, since he is largely responsible for establishing the environment for innovation within the technical departments, setting the managerial style, and introducing change where necessary. Above all he must ensure that he does not become a barrier to innovation.

8.4 Career development and competence

In the previous section the competences required at various stages in the engineer's career progression have been discussed. These are shown diagrammatically in Table 8.1, where the number of crosses indicate in broad terms the relative importance of each of these competences at the different phases. This career progression may be regarded also as two transformations: from engineer into manager, and from manager into businessman.

Whilst a few engineers remain specialists, particularly in some large companies which have established a technical career ladder, the majority become engaged in some managerial activities after a few years in industry. Although the application of these competences is associated with promotion, they can only be acquired after a number of years of learning, training and preparation. Thus the learning must be continuous, extending throughout the engineer's career.

The demands on an engineer are usually such that he may direct all his energies to succeeding in his current job, with little attention paid to preparing himself for the next phase of his career and even less to the requirements of the phase after that. Indeed he may not appreciate the need to acquire new skills. As a consequence he may well fail in his next job at worst, or muddle through at best. The result is a loss of job satisfaction and poor performance.

Of course, engineers attend many educational programmes both within the company and externally. These are usually of short duration and most of his career development must of necessity occur on the job. Companies do have a responsibility to develop their employees, but the most important contribution must come from the individual himself. In doing this he must:

1 Understand his development needs
2 Acquire knowledge of managerial and business concepts and techniques
3 Learn from his experience in applying these concepts and techniques in his own job
4 Continually update his knowledge by attending courses and reading the business and managerial literature.

This book has raised a number of issues in respect of the business as a whole which the engineer must understand if he is to succeed as his career progresses. It has discussed the most important concepts and described briefly some of the techniques. It is only a beginning for what must be a lifelong study if the engineer is to progress for the good of his own career and of the organisations which employ him.

Bibliography

This bibliography gives a selection of books for further reading. It is not intended to be comprehensive. The books do, however, explore in greater depths many of the issues raised in this text.

ACKOFF, R. L., *Creating the Corporate Future*, Wiley 1981
ALLEN, T. J., *Managing the Flow of Technology*, MIT 1977
ANSOFF, H. I., *Implanting Strategic Information*, Prentice-Hall 1984
ANSOFF, H. I., *Corporate Strategy*, Penguin 1976
BAKER, M. J., *Marketing Strategy and Management*, Macmillan 1985
BAKER, M. J., *Market Development: A Comprehensive Survey*, Penguin 1983
BAYLISS, J. S., *Marketing for Engineers*, Peter Peregrinus 1985
BROWN, J. J. and HOWARD, L. R., *Managerial Accounting and Finance*, 4th edn, Macdonald & Evans 1982
DANIEL, W. W., *Workplace Industrial Relations and Technical Change*, Frances Pinter 1987
DEAN, B. V. and GOLDHAR, J. L., *Management of Research and Innovation*, North-Holland 1980
DRUCKER, P. F., *Managing in Turbulent Times*, Heinemann 1980
FOSTER, R., *Innovation: The Attacker's Advantage*, Macmillan 1986
HIGGINS, J. C., *Strategic and Operational Planning Systems*, Prentice-Hall 1980
HIGGINS, J. C., *Information Systems for Planning and Control*, Edward Arnold 1976
HILL, T. J., *Manufacturing Strategy*, Macmillan Educational 1985
HILL, T. J., *Production/Operations Management*, Prentice-Hall 1983
HOFER, C. W. and SCHENDEL, D., *Strategy Formulation: Analytical Concepts*, West 1978
JONES, H. and TWISS, B.C., *Forecasting Technology for Planning Decisions*, Macmillan 1978
KOTLER, P., *Marketing Management: Analysis, Planning and Control*, 5th edn, Prentice-Hall 1984
LOCKYER, K., *Production Management*, 4th edn, Pitman 1983
MARTINO, J. P., *Technological Forecasting for Decision Making*, 2nd edn, North-Holland 1983
MENSCH, G., *Stalemate in Technology: Innovations Overcome the Depression*, Ballinger 1979
NORMAN, R., *Management for Growth*, Wiley 1977
PARKER, R. C., *The Management of Innovation*, Wiley 1982
PARKER, R. C., *Going for Growth: Technological Innovation in Manufacturing Industries*, Wiley 1985
PETERS, T. J. and WATERMAN, R. H., *In Search of Excellence*, Harper & Row 1982
POLICY STUDIES INSTITUTE, *Robots in British Industry: Expectations and Experience*, research report 660, 1986
PORTER, M., *Competitive Strategy*, Free Press 1980
RAY, G. F., *The Diffusion of Mature Technologies*, Cambridge University Press 1984

Bibliography

RUSKIN, A. M. and ESTES, W. E., *What Every Engineer Should Know About Project Management*, Marcel Dekker 1982
SAHAL, DAVENDRA, *Patterns of Technological Innovation*, Addison-Wesley 1981
STALLWORTHY, E. A. and KHARBANDA, O. P., *Total Project Management*, Gower 1983
TOFFLER, A., *The Third Wave*, Morrow 1980
TUSHMAN, M. L. and MOORE, W. L., *Readings in the Management of Innovation*, Pitman 1982
TWISS, B. C., *Managing Technological Innovation*, 3rd edn, Pitman 1986
TWISS, B. C. (ed.), *Managerial Implications of Microelectronics*, Macmillan 1981
TWISS, B. C. and WEINSHALL, T. D., *Managing Industrial Organizations*, Pitman 1980
VOSS, C. A. (ed.), *Managing Advanced Manufacturing Technology*, Kempston 1986
WESTON, J. F. and BRIGHAM, E. F., *Managerial Finance*, British edition, Holt, Rinehart & Winston 1979

Index

Accountants 130–3
Acquisition 14, 31
Advantage – competitive 18, 29, 47–8, 61
 – differential 18
Advocacy 194
Alaska pipeline 135
AMS 106, 124
Analysis – attribute 87–9
 – business 9–10, 51
 – competitive 51
 – cost:benefit 38, 122, 137
 – critical path 109
 – environmental 5 (see also under Forecasting)
 – gap 12–15, 51
 – market 93
 – portfolio 5, 15–17, 51
 – risk 51, 139–41, 188
 – sensitivity 64
 – strategic 6–21
 – SWOT 8–9
 – value 98
Assumptions 8, 62–63, 94–5, 195
Attitudes – to innovation 15, 156–60
 – to risk 192–3
 – senior management 81
Attributes – analysis 87–9
 – product 34, 54, 58, 60, 83–4
 – substitution 74–5
Audit – post 136, 159
 – technical 49, 51, 177

Batch size 103, 113, 121
Biotechnology 7, 162
Boston Consulting Group (BCG) 5, 15–17
Bottom-up management 41, 50, 56
Brain scanner 46

Breakeven chart 147–9
Budget – capital 133–7
 – engineering 26–7, 40, 132–3
Business ventures 5, 170–1

CAD/CAM 106, 121, 125, 184
Capabilities, technical 49
Capital – budgetting 133–7
 – intensity 148–9
Capture, technology 41, 43, 54, 82–3, 173
Career 192–3
 – development 127, 196–202
Cash – cows 16
 – flow 16, 20, 38, 71, 100–1, 131, 142–4, 194
 – management 5
Champion 14, 155–6, 159
Change – business 181–6
 – continuous 112–3
 – environmental 4
 – fear of 158–9
 – human barriers to 156–60
 – incremental 26, 46
 – management of 3, 118, 121, 129, 154–173
 – pace of 183–4
 – stimulus for 63
 – support for 166–8
Check list 94
CIM 106, 184
Collaboration 174
Communication 161, 162, 165, 169, 193–4
 – international 174–5
Competitive – advantage 18, 29, 31, 47–8, 61, 182
 – position 15–6
 – strategy 17–19

Competence 198, 201–2
Complexity 163, 187
Concorde 134
Constraints, self-imposed 160
Contingency 62, 110, 135
Contract research 43, 178
Contracts 109
Control – project 136
 – systems 33, 36
Corporate – culture 2, 14, 25, 156, 192
 – planning 4
 – strategy 1–25, 41–42
Cost 2, 85, 97
 – and pricing 100
 – benefit 38, 122, 137
 – fixed and variable 147–8
 – plus 101
 – reduction 33, 87, 97–8, 143
 – sunk 71, 144–6
Creativity 3, 19, 41, 80–2, 156, 158
Critical path analysis 109
Culture 2, 14, 25, 156, 192
Curves
 – Experience 46, 102–3, 113, 149–50
 – Gompertz 69
 – logistic 69
 – S 67–72

Data 63–5
Discounted Cash Flow (DCF) 40, 138–9
Debt capacity 4
Decisions 6, 11–2, 22
 – business 69
 – complexity 24
 – engineering 186–195
 – making 186–194
 – technical 24
 – quality of 184
Depreciation 146–7
Depth, technical 49–50
Design – and the operating system 98
 – dominant 32
 – for manufacture 96–7
 – for market 98–9
 – for minimum cost 97–8
 – product 95–9
Development 7 (see also under R&D)
Discontinuities 62
Discretion 197
Diversification 7, 14
 – design 32
Dominant – technology 49

Economy of scale 5

Engineer – career 192–3, 196–202
 – role of 23–4, 53, 182, 192, 196–201
Engineering – budget 26–7, 40, 132–3
 – decisions 186–195
 – director 27, 199–201
Entrepreneurship 5, 21–2, 75, 155, 189–90
Environment (see also under forecasting) 3–6, 34
 – scanning 10–11
Estimates 134–6
Ethics 191
Exchange agreements 174, 177, 178
Experience curve 46, 102–3, 113, 149–50
Exploratory R&D 27, 42, 90–2
Evaluation – critical 158
 – financial 137–142
 – of technical change 160–5
 – project 92–5, 155–6
 – techniques 94

Feasibility studies 27, 90
Finance director 130–1
Financial evaluation 137–142
Fisher-Pry 75–7
FMS 106, 121
Forecasting 4, 5, 7, 10–11, 13, 193
 – elements of 59–62
 – environmental 84–5
 – inputs to 62–5
 – models 63, 65–77
 – normative 78
 – production 114
 – technology 46, 51, 59–78, 184

Gap – analysis 12–15, 51
 – generation 157–8
Gatekeeper 82, 174, 179
Growth – organic 14
 – v survival 1, 2, 187

History, of technology 55–8
Horizontal integration 14, 31, 150
Hurdle rate 138

Ideas (see also under creativity)
 – evaluation of 161–2
Importance 188–9
Indicators, social 64
Industrial revolution 56
Influence diagram 66–7
Information – market 83–4
 – management of 123
 – quantitative v qualitative 194

- sources of 82–3
- systems 121, 126, 136, 142, 159
- technology (IT) 121, 123, 127, 183, 185

Infrastructure 71
Innovation 5, 31, 190
- by invasion 34, 35, 44
- need for 13
- process of 155–6

'In Search of Excellence' 11 2, 22
Integration – horizontal and vertical 14, 31, 150
- inter-functional 96, 100, 105, 126, 184–5, 191–2
- of technologies 186
- organisational 123–4
- with suppliers 124–5

Intelligence, technical 46
Interpersonal skills 110, 117–8, 198
Internationalisation – of business 182–3
- of markets 115–6
- of production 176
- of R&D 176–7
- of technology 173

Inventory 108, 114, 171
Investment, technical 38, 39

Joint ventures 162, 178
Judgement 3, 10, 14, 20, 55, 67, 132, 140, 146, 188, 195
Just-in time 124–5

Kondratieff 57

Labour intensity 7, 48
Learning curve (see under experience)
Licensing 43, 178
Life cycle 31–8, 51, 87
Life, in use 34, 142, 184
Limits, technological 35
Literacy 183–4
Long wave, the 57

Mandatory requirements 43
Maintenance 38, 98
Make or buy 150–1
Management – bottom-up 41, 50, 56
- by objectives 4
- development 199–201
- of change 3, 118, 121, 129, 154–173
- matrix 124, 172–3
- participative 50
- strategic 1–25
- style 35, 190

- systems 189–90
- top 13

Manufacture (see also under production)
- investment in 26, 28–30, 86

Market – knowledge 83–4
- needs 53
- pull 54, 80
- research 33, 70, 88
- saturation 66, 71–2
- segments 33, 34, 47, 74, 77, 86, 87–9
- share 15, 77, 103, 142, 150
- substitution 75–7

Marlow Foods 162
Matrix management 124, 172–3
Maturity 33, 36
Mensch G 57
Microelectronics 5, 9, 25, 57, 96, 121
Models – causal 65–8
- forecasting 63
- technical 91–2

Monitoring 47, 54, 59–60, 63, 179, 186
MSC 121

Net Present Value (NPV) 139
New Venture Group 170–1
Normative – forecasting 78
- goals 91

Objectives – business 1, 6, 8, 124
- hierarchy of 191
- management by 4
- personal 190–2
- production 107–8

Objective tree 90–1
Oil crisis 4, 5
Operational research (OR) 4
Organisation – heirarchical 169–70
- matrix 172–3
- new venture 170–1
- structure 122, 168–173

Over-design 86

Parameter, technical 61, 64
Participative management 50
Patents 43, 104
Patterns of progress 61, 68–72
Payback 5, 127
PERT 109
Peters T. J. and Waterman R. H. 11–12, 22
Planning – contingency 62, 110, 135
- corporate 4
- production 114–7

Porter M 17–18

Index

Portfolio 12, 100, 135–6
 – analysis 5, 15–17, 51
Pre-competitive research 44, 174
Pressures 19–21
Price 32, 101–4
 – elasticity 101
Priority 188–9
Proactive 3, 4, 24
Probability 61 164–5
Product – attributes 34, 54, 58, 60, 83–4
 – concept 80–5
 – design 95–9
 – development 104–5
 – life 71
 – longevity 34
 – maturity 33, 36
 – performance 71, 85–6
 – specification 85–90
 – substitution 75–7, 183
 – variety 33, 108
Production 106–128
 – and R&D 125
 – manager 126–7
 – mass 113
 – one-off 108–112
 – planner 114–7
 – process 113, 116, 118–9
 – repetitive 113–119
 – role of 107–8
Productivity 133
 – of R&D 72, 75, 87
Profile, risk 50
Profitability 2, 12, 100, 131, 194
Project – champion 94, 155–6, 169
 – control 136–7
 – evaluation 92–5
 – management 108–12, 172–3, 185, 199
 – proposals 50–1

Qualitative factors 59, 187–8
Quality – assurance 128–9
 – circles 126
Quantification 4, 60, 167, 188, 194

Reactive management 3
Relevance tree 78, 90–1
Research – applied 7, 85, 91
 – contract 43, 178
 – pre-competitive 44, 174
Research & Development (R&D) 15, 23, 26
 – and production 125
 – cost of 173–4
 – exploratory 27, 42, 90–2
 – financial estimates 134–5
 – international 176–7
 – investment 33, 134, 146
 – management 198
 – productivity 72, 75, 87
 – siting 176–8
 – strategy 27–8
 – values 191
Return on Investment (ROI) 137–8, 142–3
Risk 1, 2, 8, 50, 92, 164–5, 192–3

Saturation, market 68
Scanning 10, 35
Scenarios 5, 22, 78, 193
Science 53
Scientific method 193
S-curve 67–72
Section leader 197
Security 192–3
Seikan tunnel 135
Shareholders 130–1
Silicon Valley 177
Social – indicators 64
 – motives 2
 – trends 66
Specialist 197
Specification – product 85–90
 – technical 61
Stakeholders 21
Status 32–3, 162, 194
 – review 39–41
Stocks (see under inventory)
Strategy – corporate 1–25
 – defensive 47
 – formulation 8
 – financial 42
 – market 42
 – offensive 47
 – technical 26–52
Substitution – attribute 74–5
 – product 75–7, 183
 – technological 72–4
Suppliers 124–5
Survival 1, 2, 187
SWOT analysis 8–9
Systems – compatability 98
 – control 33, 36
 – total 61

Technical – audit 49, 51, 177
 – capability 49
 – change programme 160–5
 – director 27, 199–201

– exploration 42
– excellence 191
– intelligence 46
– investment 38–9
– knowledge 196–7
– parameter 61, 64
– progress 58–9
– specification 61
– strategy 26–52
Technological limits 35
Technology – capture 41, 43, 82–3, 173
– competence 173
– forecasting (TF) (see under forecasting)
– push 54, 80
– transfer 173–9
Time 60–1, 85
– development 96
– lead 70, 114
Timing 46–7, 75
Top-down management 41, 156
Training 193, 198

Uncertainty 1, 2, 62, 94, 134–5, 139–41, 159, 163, 183, 192–3

Urgency 143–4, 184, 189
User – ideas 84
– needs 95

Value – added 48, 151–2, 182
Values – corporate 2, 12, 191
– personal 191–2
– traditional 27
Variety 33, 108
Variance 132
Ventures
– business 5, 170–1
– joint 162, 178
– new 57, 170–1
Vertical integration 31, 151
Voss C.A. 125
Vulnerability 31, 32, 183

Workforce 126

Yamazaki 121

Zero defects 128